SEVENTH EDITION

Microbiology Experiments

A HEALTH SCIENCE PERSPECTIVE

John Kleyn

Mary Bicknell
University of Washington

Anna Oller
University of Central Missouri

Mc
Graw
Hill

*Connect
Learn
Succeed*™

MICROBIOLOGY EXPERIMENTS: A HEALTH SCIENCE PERSPECTIVE, SEVENTH EDITION

Published by McGraw-Hill, a business unit of The McGraw-Hill Companies, Inc., 1221 Avenue of the Americas, New York, NY 10020. Copyright © 2012 by The McGraw-Hill Companies, Inc. All rights reserved. Printed in the United States of America. Previous editions © 2009, 2007, and 2004. No part of this publication may be reproduced or distributed in any form or by any means, or stored in a database or retrieval system, without the prior written consent of The McGraw-Hill Companies, Inc., including, but not limited to, in any network or other electronic storage or transmission, or broadcast for distance learning.

Some ancillaries, including electronic and print components, may not be available to customers outside the United States.

 This book is printed on recycled, acid-free paper containing 10% postconsumer waste.

3 4 5 6 7 8 9 0 QVS/QVS 1 0 9 8 7 6 5 4

ISBN 978–0–07–731554–2
MHID 0–07–731554–5

Vice President, Editor-in-Chief: *Marty Lange*
Vice President, EDP: *Kimberly Meriwether David*
Senior Director of Development: *Kristine Tibbetts*
Publisher: *Michael S. Hackett*
Sponsoring Editor: *Lynn M. Breithaupt*
Senior Developmental Editor: *Fran Simon*
Director of Digital Content Development: *Barbekka Hurtt, PhD*
Marketing Manager: *Amy L. Reed*
Project Manager: *Mary Jane Lampe*
Senior Buyer: *Laura Fuller*
Designer: *Tara McDermott*
Cover/Interior Designer: *Elise Lansdon*
Cover Image: *Enterococcus faecalis Bacteria. SEM X3000. ©Dennis Kunkel Microscopy, Inc./Visuals Unlimited, Inc.*
Senior Photo Research Coordinator: *John C. Leland*
Photo Research: *David Tietz/Editorial Image, LLC*
Compositor: *S4Carlisle Publishing Services*
Typeface: *11.5/13 Goudy Old Style*
Printer: *Quad/Graphics*

All credits appearing on page or at the end of the book are considered to be an extension of the copyright page.

Some of the laboratory experiments included in this text may be hazardous if materials are handled improperly or if procedures are conducted incorrectly. Safety precautions are necessary when you are working with chemicals, glass test tubes, hot water baths, sharp instruments, and the like, or for any procedures that generally require caution. Your school may have set regulations regarding safety procedures that your instructor will explain to you. Should you have any problems with materials or procedures, please ask your instructor for help.

www.mhhe.com

CONTENTS

PREFACE

To the Student

A microbiology laboratory is valuable because it actually gives you a chance to see and study microorganisms firsthand. In addition, it provides you with the opportunity to learn the special techniques used to study and identify these organisms. The ability to make observations, record data, and analyze results is useful throughout life.

It is very important to read the scheduled exercises before coming to class so that class time can be used efficiently. It is helpful to ask yourself the purpose of each step as you are reading and carrying out the steps of the experiment. Sometimes it will be necessary to read an exercise several times before it makes sense.

Conducting experiments in microbiology laboratories is particularly gratifying because the results can be seen in a day or two (as opposed, for instance, to plant genetics laboratories). Opening the incubator door to see how your cultures have grown and how the experiment has turned out is a pleasurable moment. We hope you will enjoy your experience with microorganisms as well as acquire skills and understanding that will be valuable in the future.

To the Instructor

The manual includes a wide range of exercises— some more difficult and time-consuming than others. Usually more than one exercise can be done in a 2-hour laboratory period. In these classes, students can actually see the applications of the principles they have learned in the lectures and text. We have tried to integrate the manual with the text *Microbiology: A Human Perspective*, Seventh Edition, by Eugene Nester et al.

The exercises were chosen to give students an opportunity to learn new techniques and to expose them to a variety of experiences and observations. We did not assume that the school or department has a large budget; thus, exercises were written to use as little expensive media and equipment as possible. The manual contains more exercises than can be done in one course so that instructors will have an opportunity to select the appropriate exercises for their particular students and class.

- An online Instructor's Manual is available from the publisher.
  - It lists equipment, cultures, media, and reagents needed for each exercise and has extensive information for storing cultures and making media.
- The Notes to the Instructor section gives suggestions for preparing and presenting the laboratory sessions.
- There are questions that can be used to supplement those in the student manual.
- **We highly recommend that the instructors utilize the Instructor's Manual.** Contact your local McGraw-Hill representative for the URL and password for this site.
- Revisions to the Seventh Edition include
  - New exercises:
    - Epidemiology
    - ELISA (simplified)
    - PCR
    - Microarray
  - Changes to exercises:
    - The Antibiotics exercise was separated from Antiseptics and Disinfectants.
    - *Propionibacterium* was included with Normal Skin Biota.
    - Streptococci was included with Respiratory Microorganisms.
  - Twenty-two new color plates
  - Three Prelab questions for each exercise
  - Media and Reagent formulas in Appendix 1
  - The metric system with conversions on the inside back cover

- Some of the exercises have been condensed and rearranged to give a more logical introduction to techniques, and others have been simplified.
- If a student has a special interest in microbiology and would like to do independent work, four special projects are included, of which two projects are available online:
  o The isolation of *Deinococcus*, an organism exceptionally resistant to radiation
  o The isolation of an organism utilizing methanol from plant leaves
- One exercise, Conjugation, is available online.
- You can access the exercise and special projects at www.mhhe.com/nester7. Isolation techniques for two other interesting organisms, petroleum-utilizing bacteria (important in oil-polluted water) and bioluminescent bacteria, are in the *back of the manual*. The luminescent bacteria are particularly intriguing. Not only are they an example of quorum sensing, but they are exciting to view in a dark room. Also, these bacteria can be easily isolated by the whole class.

We hope these changes are helpful and that the manual contributes to the students' understanding of microbiology. We also hope both students and instructors enjoy their experience with a very interesting group of organisms.

Acknowledgments

We would like to acknowledge the contributions of the lecturers in the Department of Microbiology at the University of Washington who have thoughtfully honed laboratory exercises over the years until they really work. These include Carol Laxson, Mona Memmer, Janis Fulton, and Mark Chandler.

Special thanks to Dale Parkhurst for his expert knowledge of media. We also thank the staff of the University of Washington media room, headed by Jim Calbreath, and William Kirby at the University of Central Missouri for their expertise and unstinting support.

We also want to thank Eugene Nester, Denise Anderson, and Evans Roberts for their text *Microbiology: A Human Perspective*. This text was the source of much of the basic conceptual material and figures for our laboratory manual.

We gratefully acknowledge the contributions of Suzanne Schlador, from North Seattle Community College. Her ideas and explanations of current techniques were very valuable in our efforts to update this manual.

We also thank the following instructors for their valuable input on the revision of this manual.

Reviewers

Barbara N. Beck
Rochester Community and Technical College

Mary Anne Frazee
Urbana University

Ilko Iliev
Southern University at Shreveport

Subburaj Kannan
Southwest Texas Junior College

Kristy McClain
East Mississippi Community College

Dr. Chere Mosley
Ivy Tech Community College, Bloomington

Sara Reed Houser
Jefferson College of Health Sciences

Ken E. Roth
James Madison University

Robert Walters
James Madison University

We would also like to express our appreciation to Fran Simon, Lynn Breithraupt, Mary Jane Lampe, and all the others involved in publishing this manual.

LABORATORY SAFETY

To be read by the student before beginning any laboratory work.

1. Do not eat, drink, smoke, or store food in the laboratory. Avoid all finger-to-mouth contact. Do not chew on pens or pencils.
2. Never pipet by mouth because of the danger of ingesting microorganisms or toxic chemicals.
3. Wear a laboratory coat while in the laboratory. Remove it before leaving the room and store it in the laboratory until the end of the course.*
4. Wipe down the bench surface with disinfectant before and after each laboratory period.
5. Tie long hair back to prevent it from catching fire in the Bunsen burner or contaminating cultures.
6. Keep the workbench clear of any unnecessary books or other items. Do not work on top of the manual because if spills occur, it cannot be disinfected easily.
7. Be careful with the Bunsen burner. Make sure that paper, alcohol, the gas hose, and your microscope are not close to the flame.
8. All contaminated material and cultures must be placed in the proper containers for autoclaving before disposal or washing.
9. Mix cultures gently to avoid creating aerosols. Clean off the loop in a sand jar before flaming in the Bunsen burner.
10. If a culture is dropped and broken, notify the instructor. Cover the contaminated area with a paper towel and pour disinfectant over the material. After 10 minutes, put the material in a broken-glass container to be autoclaved.
11. Carefully follow the techniques of handling cultures as demonstrated by the instructor.
12. When the laboratory is in session, keep the doors and windows shut. Post a sign on the door indicating that a microbiology laboratory is in session.
13. Be sure you know the location of fire extinguishers, eyewash apparatus, and other safety equipment.
14. Wash your hands with soap and water after any possible contamination and at the end of the laboratory period.
15. If you are immunocompromised for any reason (including pregnancy), consult a physician before taking this class.

---

*Other protective clothing includes closed shoes, gloves (optional), and eye protection.

LABORATORY SAFETY AGREEMENT

To be read by the student before beginning any laboratory work.

1. Do not eat, drink, smoke, or store food in the laboratory. Avoid all finger-to-mouth contact. Do not chew on pens or pencils.
2. Never pipet by mouth because of the danger of ingesting microorganisms or toxic chemicals.
3. Wear a laboratory coat while in the laboratory. Remove it before leaving the room and store it in the laboratory until the end of the course.*
4. Wipe down the bench surface with disinfectant before and after each laboratory period.
5. Tie long hair back to prevent it from catching fire in the Bunsen burner** or contaminating cultures.
6. Keep the workbench clear of any unnecessary books or other items. Do not work on top of the manual because if spills occur, it cannot be disinfected easily.
7. Be careful with the Bunsen burner.** Make sure that paper, alcohol, the gas hose, and your microscope are not close to the flame.
8. All contaminated material and cultures must be placed in the proper containers for autoclaving before disposal or washing.
9. Mix cultures gently to avoid creating aerosols. Clean off the loop in a sand jar before flaming in the Bunsen burner.
10. If a culture is dropped and broken, notify the instructor. Cover the contaminated area with a paper towel and pour disinfectant over the material. After 10 minutes, put the material in a broken glass container to be autoclaved.
11. Carefully follow the techniques of handling cultures as demonstrated by the instructor.
12. When the laboratory is in session, keep the doors and windows shut. Post a sign on the door indicating that a microbiology laboratory is in session.
13. Be sure you know the location of fire extinguishers, eyewash apparatus, and other safety equipment.
14. Wash your hands with soap and water after any possible contamination and at the end of the laboratory period.
15. If you are immunocompromised for any reason (including pregnancy), consult a physician before taking this class.

---

* Other protective clothing includes closed shoes, gloves (optional), and eye protection.

** **Note to Instructor:** Modify these rules if you do not use a Bunsen burner.

I have read and understood the laboratory safety rules:

_____   _____
                         Signature   Date

INTRODUCTION to Microbiology

When you take a microbiology class, you have an opportunity to explore an extremely small biological world that exists unseen in our own ordinary world. Fortunately, we were born after the microscope was perfected, so we can observe these extremely small organisms. At this time, we also actually understand how the cell transfers the information in its genes to the operation of its cellular mechanisms so that we can study not only what bacteria do but also how they do it.

A few of these many and varied organisms are pathogens (capable of causing disease). Special techniques have been developed to isolate and identify them as well as to control or prevent their growth. The exercises in this manual will emphasize medical applications. The goal is to teach you basic techniques and concepts that will be useful to you now or can be used as a foundation for additional courses. In addition, these exercises are also designed to help you understand basic biological concepts that are the foundation for applications in all life science fields.

As you study microbiology, it is also important to appreciate the essential contributions of microorganisms as well as their ability to cause disease. Most organisms play indispensable roles in breaking down dead plant and animal material into basic substances that can be used by other growing plants and animals. Photosynthetic bacteria are an important source of the earth's supply of oxygen. Microorganisms also make major contributions in the fields of antibiotic production, food and beverage production, as well as food preservation and, more recently, recombinant DNA technology. The principles and techniques demonstrated here can be applied to these fields as well as to medical technology, nursing, or patient care. This course is an introduction to the microbial world, and we hope you will find it useful and interesting.

In the next few exercises, you will be introduced to several basic procedures: the use of the microscope and aseptic and pure culture techniques. These are skills that you will use not only throughout the course but in any microbiology laboratory work you do in the future.

Note: The use of pathogenic organisms has been avoided whenever possible, and nonpathogens have been used to illustrate the kinds of tests and procedures that are actually carried out in clinical laboratories. In some cases, however, it is difficult to find a substitute, and organisms of low pathogenicity are used. These exercises will have an additional safety precaution.

Notes:

warm blooded - 37°
soil organisms - 30°C
~~greater organisms~~ - 25°C
room surfaces

EXERCISE 1

Ubiquity of Microorganisms

Getting Started

Microorganisms are everywhere—in the air, soil, and water; on plant and rock surfaces; and even in such unlikely places as Yellowstone hot springs and Antarctic ice. Millions of microorganisms are also found living within or on animals; for example, the mouth, the skin, and the intestine (all exterior to our actual tissues) support huge populations of bacteria. In fact, the interior of healthy plant and animal tissues is one of the few places free of microorganisms. In this exercise, you will sample material from the surroundings and your body to determine what organisms are present that will grow on laboratory **media.**

An important point to remember as you try to grow organisms is that there is no one condition or **medium** that will permit the growth of all microorganisms. The trypticase soy **agar** used in this exercise is a rich medium (a digest of meat and soy products, similar to a beef and vegetable broth) and will support the growth of many diverse organisms, but bacteria growing in a freshwater lake that is very low in organic compounds would find it too rich (similar to a goldfish in vegetable soup). However, organisms that are accustomed to living in our nutrient-rich throat might find the same medium lacking substances they require.

Temperature is also important. Organisms associated with warm-blooded animals usually prefer temperatures close to 37°C, which is approximately the body temperature of most animals. Soil organisms generally prefer a cooler temperature of 30°C. Organisms growing on glaciers would find room temperature (about 25°C) much too warm and would probably grow better in the refrigerator.

Microorganisms also need the correct atmosphere. Many bacteria require oxygen, whereas other organisms find it extremely toxic and will only grow in the absence of air. Therefore, the organisms you see growing on the plates may be only a small sample of the organisms originally present.

Definitions

Agar. A carbohydrate derived from seaweed used to solidify a liquid medium.

Colony. A visible population of microorganisms growing on a solid medium.

Inoculate. To transfer organisms to a medium to initiate growth.

Media (singular, *medium*). The substances used to support the growth of microorganisms.

Normal biota. The organisms usually found associated with parts of the body (previously termed *normal flora*).

Pathogen. An organism capable of causing disease.

Sterile. Free of viable microorganisms and viruses.

Ubiquity. The existence of something everywhere at the same time.

Objectives

1. Explain that organisms are **ubiquitous.**
2. Explain how organisms are grown on laboratory culture **media.**

Prelab Questions

1. Why isn't there one kind of agar medium that will support the growth of all microorganisms?
2. We require oxygen to live. Is this also true of all bacteria?
3. What is *normal biota?*

Materials

Per team of two (or each individual, depending on number of plates available):

Trypticase soy agar plates, 2
Sterile swabs as needed
Sterile water (about 1 ml/tube) as needed
Waterproof marking pen or wax pencil

PROCEDURE

First Session

1. Each pair of students should obtain two petri plates of trypticase soy agar. Notice that the lid of a petri plate fits loosely over the bottom half.

2. Label the plates with your name and date using a wax pencil or waterproof marker. Always label the bottom of the plate because sometimes you may be examining many plates at the same time and it is easy to switch the lids.

3. Divide each plate in quarters with two lines on the bottom of the petri plate. Label one plate 37°C and the other 25°C (figure 1.1).

4. **Inoculate** the 37°C plate with samples from your body. For example, moisten a sterile swab with sterile water and rub it on your skin and then on one of the quadrants (be careful not to gouge the agar). Try touching your fingers to the agar before and after washing, or place a hair on the plate. Try whatever interests you. (Be sure to place all used swabs into an autoclave container or bucket of disinfectant after use.)

5. Inoculate the plate labeled 25°C (room temperature) with samples from the room. It is easier to pick up a sample if the swab is moistened in sterile water first. **Sterile** water is water in which there are no living organisms and is used so that your results will not be contaminated. Try sampling the bottom of your shoe or some dust, or press a coin or other objects lightly on the agar. Be sure to label each quadrant so that you will know what you used as inoculum.

6. Incubate the plates at the temperature written on the plate. Place the plates in the incubator or basket upside down. This is important because it prevents condensation from forming on the lid and dripping on the agar below. The added moisture would permit colonies of bacteria to run together.

Second Session

Important: Handle all plates with colonies as if they were potential **pathogens,** disease-causing organisms. Follow your instructor's directions carefully.

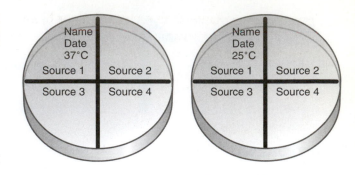

Figure 1.1 Plates labeled on the bottom for ubiquity exercise.

Note: For best results, the plates incubated at 37°C should be observed after 2 days, but the plates at room temperature will be more interesting at about 5–7 days. If possible, place the 37°C plates either in the refrigerator or at room temperature after 2 days so that all the plates can be observed at the same time.

1. Examine the plates you prepared in the first session and record your observations on the report sheet for this exercise. There will be basically two kinds of **colonies:** fungi (molds) and bacteria. Mold colonies are usually large and fluffy, the type found on spoiled bread. Bacterial colonies are usually soft and glistening and tend to be cream colored or yellow. Compare your colonies with color plates 1 and 2.

2. When describing the colonies include

   a. relative size as compared to other colonies

   b. shape (round or irregular)

   c. color

   d. surface (shiny or dull)

   e. consistency (dry, moist, or mucoid)

   f. elevation (flat, craterlike, or conical)

3. There may be surprising results. If you pressed your fingers to the agar before and after washing, you may find more organisms on the plate after you washed your hands. The explanation is that your skin has **normal biota** (organisms that are always found growing on your skin). When you wash your hands, you wash off the organisms you have picked up from your surroundings as well as a few layers of skin. This exposes more of your normal biota; therefore, you may see different colonies of bacteria before

you wash your hands than afterward. Your biota is important in preventing undesirable organisms from growing on your skin. Hand washing is an excellent method for removing pathogens that are not part of your normal biota.

4. (Optional) If desired, use these plates to practice making simple stains or Gram stains as described in exercises 5 and 6.

Note: In some labs, plates with molds are opened as little as possible and immediately discarded in an autoclave container to prevent contaminating the lab with mold spores.

5. Follow the instructor's directions for discarding plates. All agar plates are autoclaved before washing or discarding in the municipal garbage system.

First Session

1. 2 petri dish - trypticase soy agar.

   a. label plates

3. divide into two sections 37° and 25°C

4. inoculate 37°C with skin/body samples

5. innoculate 25°C with room samples

6. incubate upside down.
   { wait 2 days }

Second Session

1. Examine plates { there will be 2 kinds of colonies }
   (mold) fungi, bacteria - soft, glistening; white yellow

2. Describe colonies by: relative size to other colonies
   - shape
   - color
   - consistency
   - elevation

3. - normal biota (organisms always found living on the skin)

4. - optional, save plates for gram stain

5. - discard materials properly

EXERCISE

1

Laboratory Report: Ubiquity of Microorganisms

Results

Room Temperature (about 25°C) Plate

	Plate Quadrant			
	1	**2**	**3**	**4**
Source				
Colony appearance				

37°C Plate

	Plate Quadrant			
	1	**2**	**3**	**4**
Source				
Colony appearance				

Questions

1. Give three reasons why all the organisms you placed on the trypticase soy agar plates might not grow.

2. Why were some agar plates incubated at 37°C and others at room temperature?

3. Why do you invert agar plates when placing them in the incubator?

4. Name one place that might be free of microorganisms.

EXERCISE

2

Pure Culture and Aseptic Technique

Aseptic Technique and Streak Plate Technique

Getting Started

The two goals of **aseptic** technique are to prevent contamination of your culture with organisms from the environment and to prevent the culture from contaminating you or others. In this exercise you will learn three important procedures using aseptic technique: transferring material with a sterile flamed loop, transferring liquid with a sterile pipet, and isolating a pure culture.

First you will transfer sterile broth back and forth from one tube to another using a sterile flamed loop. The goal is to do it in such a way that no organism in the environnment enters the tubes. Then you will use a sterile pipet to aseptically transfer broth between the same tubes of broth. After you have practiced transferring the broth, you will **incubate** the broth tubes for a few days to determine if they are still sterile. If you used good technique, the broth will still be clear; if organisms were able to enter from the environment, the broth will be cloudy from the bacterial growth. When you can successfully transfer **sterile** broth aseptically, you can use the same technique to transfer a pure culture without contaminating it or the environment.

A control broth will also be incubated with the practice tubes. This tube is inoculated with *Escherichia coli* to show that the broth supports bacterial growth. Without this control you would not know if the practice tubes were clear after incubation because of excellent technique or because bacteria could not grow in the broth.

The **streak plate** is the third essential technique. It permits you to isolate a colony formed by a single cell from a mixture containing millions of cells. You will start with a broth culture containing two different organisms and separate them into two different colony types. By spreading the bacteria around on the surface of the plate, the cells are able to form individual colonies separated from other individual colonies (color plate 3). This technique is used to obtain a **pure culture**.

Definitions

Aseptic. Free of contamination.

Incubate. Store cultures under conditions suitable for growth, often in an incubator.

Pure culture. A population of cells resulting from the growth of a single cell.

Sterile. Free of viable bacteria or viruses.

Streak Plate. A technique for isolating pure cultures by spreading organisms on an agar plate.

Objectives

1. Describe aseptic technique procedures.
2. Describe the isolation of separate colonies using the streak plate technique.

Prelab Questions

1. What is the purpose of transferring sterile broth from one tube to another tube of sterile broth?
2. How will you be able to determine if bacteria entered the broth during the transfers?
3. How do make sure that there are no organisms on your wire loop?

Note: Some laboratories use electric incinerators instead of Bunsen burners (figure 2.1). Others have neither heat source and perform all transfers with disposable sterile loops. Follow directions of your instructor for appropriate procedures.

Aseptic Technique

Broth-to-Broth Transfer with a Wire Loop

Materials

Per student
   Tubes of Trypticase soy broth, 3
   Inoculating loop
   Broth culture of *Escherichia coli*

Figure 2.1 An electric incinerator used to sterilize loops and needles. Courtesy of Anna Oller, Univerity of Central Missouri.

Procedure

1. Always label tubes before adding anything to them. In this exercise, you will be transferring sterile broth from one tube to another, so two tubes can be labeled "P," for practice. Label the third tube "C," for control. Always include the date on anything you label.

2. Grip the loop as you would a pencil and flame the wire portion red hot. Hold it at an angle so that you do not burn your hand (figure 2.2).

3. After the loop has cooled for a few seconds, pick up a practice tube in the other hand and remove the cap of the tube with the little finger (or the fourth and little fingers) of the hand holding the loop.

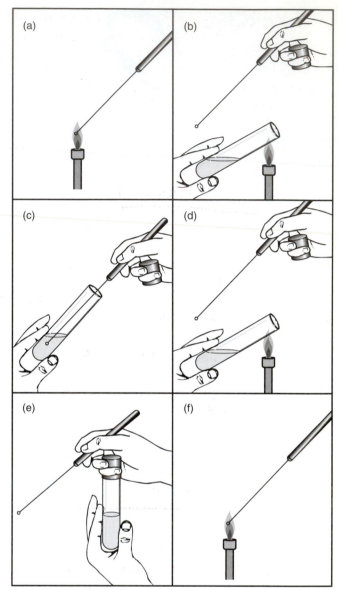

Figure 2.2 Aseptic technique for removing a loopful of broth culture. (*a*) Hold the culture tube in your left hand and the loop in your right hand (reverse if you are left-handed). Flame the loop to sterilize it. (*b*) Remove the culture tube plug or cap, and flame the mouth of the culture tube. (*c*) Insert the sterile loop into the culture tube. (*d*) Remove the loopful of inoculum, and flame the mouth of the culture tube again. (*e*) Replace the culture tube plug or cap. Place the culture tube in a test tube rack. (*f*) Reflame the loop.

4. Flame the mouth of the tube by passing it through a Bunsen burner flame, and then use the sterile loop to obtain a loopful of liquid from the tube. Flame the mouth of the tube again and replace the cap. If you have trouble picking up a loopful of material, check to be sure that your loop is a complete circle without a gap.

5. Set down the first tube and pick up the second practice tube. Remove the cap, flame it, and deposit a loopful of material into the liquid of the second tube. Withdraw the loop, flame the tube, and then replace the cap. Be sure to flame the loop before setting it on the bench (your loop would normally be contaminated with the bacteria you were inoculating). It is usually convenient to rest the hot loop on the edge of the Bunsen burner.

6. Using the loop as before, inoculate the control tube with the broth culture provided by the instructor. Incubate it with the other tubes.

7. When learning aseptic technique, it is better to hold one tube at a time; later, you will be able to hold two or three tubes at the same time.

   Note: Many laboratories no longer flame the mouth of the tube, especially if the tubes are plastic. However some laboratories still follow this procedure.

Broth-to-Broth Transfer with a Pipet

Note: Sterile pipets are used when it is necessary to transfer known amounts of material. Some laboratories use plastic disposable pipets and others use reusable glass pipets. Be sure to follow the instructor's directions for proper disposal after use (**never put a used pipet on your bench top**). Mouth pipetting is dangerous and is not permitted. A variety of bulbs or devices are used to draw the liquid up into the pipet, and your laboratory instructor will demonstrate their use (figure 2.3).

The same broth tubes used for practice with the loop can be used to practice pipetting the broth back and forth.

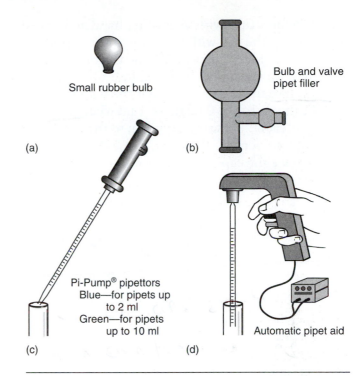

Small rubber bulb
(a)

Bulb and valve pipet filler
(b)

Pi-Pump® pipettors
Blue—for pipets up
to 2 ml
Green—for pipets
up to 10 ml
(c)

Automatic pipet aid
(d)

Figure 2.3 Various devices for filling pipets.

Materials

Per student
Trypticase Soy broth tubes from previous procedure
1 ml sterile pipet
Bulb or other device to fit on end of the pipet

Procedure

First Session

1. Open a sterile pipet at the top, insert a bulb on the end, then carefully remove the pipet from the package or canister without touching the tip. Grip the pipet as you would a pencil. The pipet is plugged with cotton to filter the air going into it. Discard the pipet if liquid inadvertently wets the plug—air will no longer enter the pipet and the measured liquid will not flow out. Notify your instructor if the bulb is contaminated.

2. Pick up a tube with your other hand and remove the cap with the little finger of the hand holding the pipet. Flame the tube. Expel air from the rubber bulb and insert the pipet tip into the liquid. Note that the liquid must be drawn to the 0 mark for 1 ml when using a 1 ml pipet. Draw the liquid up to the desired amount, remove the pipet, flame the tube, replace the cap, and then put the tube back in the rack.

3. Pick up the second tube and repeat the steps used with the first tube except that the liquid is expelled into the tube.

4. Repeat the above steps with the same tubes until you feel comfortable with the procedure.

5. Dispose of the pipet as directed.

6. Incubate the tubes at 37°C until the next period.

Second Session

1. Observe the control broth and note if it is cloudy. This cloudiness indicates growth and demonstrates that the broth can support the growth of organisms.

2. Observe the practice tubes of broth for cloudiness. Compare them to an uninoculated tube of broth. If they are cloudy, organisms contaminated the broth during your practice and grew during incubation. With a little more practice, you will have better technique. If the broths are clear, there was probably no contamination and you transferred the broth without permitting the entry of any organisms into the tubes.

3. Record results.

Streak Plate Technique

Materials

Per student

Trypticase Soy agar plates, 2

Broth culture containing a mixture of two organisms, such as *Micrococcus* and *Staphylococcus*

First Session

1. Label the agar plate on the bottom with your name and date.

2. Divide the plate into three sections with a *T* as diagrammed (figure 2.4).

3. Sterilize the loop in the flame by heating the whole length of the wire red hot. Hold it at an angle so you do not heat the handle or roast your hand.

4. Gently shake the culture to be sure both organisms are suspended. Aseptically remove a loopful of the culture and, holding the loop as you would a pencil, spread the bacteria on section 1 of the plate by streaking back and forth. The more streaks, the better chance of an isolated colony. As you work, partially cover the petri dish with the cover to minimize organisms falling on the plate from the air. Use a gliding motion and avoid digging into the agar. Don't press the loop into the surface. If your loop is not smooth or does not form a complete circle, it can gouge the agar and colonies will run together. Note that you can see the streak marks if you look carefully at the surface of the plate.

5. Burn off all the bacteria from your loop by heating it red hot. This is very important because it eliminates the bacteria on your loop. Wait a few seconds to be sure the loop is cool.

6. Without going into the broth again, streak section 2 (see figure 2.4) of the petri plate. Go into section 1 with about three streaks and spread by filling section 2 with closely spaced streaks. Be very careful not to touch the streaks in section 1 again.

7. Again heat the loop red hot. Go into section 2 with about three streaks and spread by filling section 3 with streaks. The more streaks you are able to make, the greater will be your chance of obtaining isolated colonies.

8. Heat the loop red hot before placing it on the bench top. Usually you can rest it on some part of the Bunsen burner so that it can cool without burning anything.

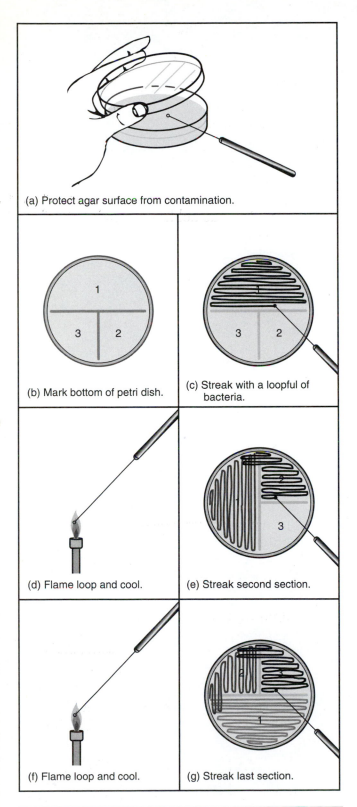

(a) Protect agar surface from contamination.

(b) Mark bottom of petri dish.

(c) Streak with a loopful of bacteria.

(d) Flame loop and cool.

(e) Streak second section.

(f) Flame loop and cool.

(g) Streak last section.

Figure 2.4 (*a–g*) Preparation of a streak plate.

9. Repeat the procedure with a second plate for additional practice.

10. Incubate the plates, inverted, in the 37°C incubator.

Second Session

Observe your streak plates and record results.

EXERCISE

2

Laboratory Report: Pure Culture and Aseptic Technique

isolation of

Results

	Clear	Turbid
Tube 1		
Tube 2		
Control		

	Number of Colonies Isolated
Streak plate 1	
Streak plate 2	

After observing the broth tubes, had you been able to transfer the broth aseptically?

After observing the control broth, were bacteria able to grow in the broth?

Did you obtain isolated colonies of each culture?

Questions

1. What is the definition of a *pure culture*?

2. Why is sterile technique important? Give two reasons.

3. What did the control broth inoculated with *Escherichia coli* demonstrate?

4. What is the purpose of a streak plate?

5. Why is it important to avoid digging into the agar with the loop?

EXERCISE

3

Introduction to the Compound Light Microscope

Getting Started

Microbiology is the study of organisms too small to be seen with the naked eye, including a vast array of bacteria, viruses, protozoa, fungi, and algae. Van Leeuwenhoek, a Dutch merchant in the late 17th century whose hobby was lens making, was the first to see these previously unknown creatures. His microscope consisted of one simple lens, but it was enough to observe some of these tiny living things (figure 3.1). Although he made drawings of some of these organisms, he did not suspect that they were essential for the existence of our world or that a small

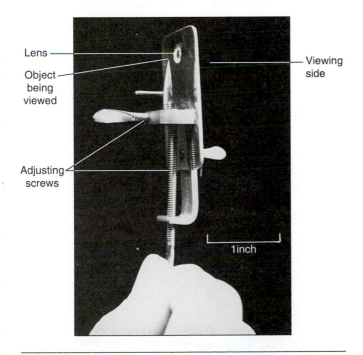

Figure 3.1 Model of a van Leeuwenhoek microscope. The original was made in 1673 and could magnify the object being viewed almost 300×. The object being viewed is brought into focus with the adjusting screws. This replica was made according to the directions given in the *American Biology Teacher* 30:537, 1958. Note its small size. Photograph courtesy of J.P. Dalmasso.

percentage was responsible for contagious diseases. It was only after the development of the compound microscope almost 200 years later that it was possible to understand the role of microorganisms in disease. After the microscope was perfected in the 1870s and 1880s, real progress was made in determining the actual cause of disease (see Appendix 6 for a history of the development of the microscope). In 1877, Robert Koch saw bacteria in the blood from an animal with anthrax, and for the first time it could be seen that bacteria, not swamp gas or evil spirits, could cause disease.

The modern compound light microscope consists of two lens systems. The first is the objective lens, which is closest to the material on the slide, and the second is the eyepiece, or ocular lens, which magnifies the image formed by the objective lens. The total magnification is found by multiplying the eyepiece lens magnification by the objective lens magnification. For example, if the eyepiece lens magnification is 10× and the objective lens magnification is 45×, the total magnification is 450 diameters.

Although it is possible to put additional magnifying glasses on top of the eyepiece lenses of the microscope, they would not improve the ability to see more detail. The reason is that the actual limiting factor of the light microscope is the resolution. This is the ability to distinguish two close objects as distinct from one another rather than as one round, hazy object. The resolution of a lens is limited by two factors: the angle of the lens and the wavelength of light entering the microscope. When using an objective lens of magnification 100× and light is optimal, a compound light microscope has a maximum magnification of about 1,000×.

A magnification of 1,000× is sufficient to easily visualize single-celled organisms such as algae and fungi. Bacteria, however, still appear very small (about the size of the letter *l* on a printed page), and their appendages, such as flagella, cannot be observed at all. Viruses are also too small to be seen in a light microscope.

To surmount the limitations of light and lens, the electron microscope, which uses electrons instead of light, was developed in the 1930s. It can magnify objects 100,000× and thus permits the visualization of viruses and structures within cells. Because the electron microscope is a very large piece of equipment requiring specialized techniques, it usually is found only in universities or research facilities. More recently, many other microscopes have been developed but usually for very specific research applications.

In this exercise, you will have an opportunity to become familiar with a compound light microscope and learn how to use and care for it. You will prepare wet mounts of unstained organisms and also learn to examine previously stained and unstained organisms. The microscope is an expensive and complex piece of equipment. Treat it with great care.

The Parts of the Microscope

The eyepieces are the lenses at the top of the microscope. They usually have a magnification of 10× (see figure 3.2).

Most microscopes have at least three objective lenses. They usually are

Low power	10×
High power	40×
Oil immersion	100×

These lenses can be rotated and the desired lens clicked into place.

The stage is below the objective lens. It usually has a device called a *mechanical stage* for holding the slide as well as knobs that permit the slide to be moved smoothly while viewing.

Below the stage is the **condenser,** which focuses the light on the slide. If it is lowered, the amount of light is reduced, but the resolution is also lowered. For our purposes, the condenser should remain at its highest position under the stage.

A lever on top of the condenser but under the stage controls the **iris diaphragm.** The diaphragm is important for adjusting the amount of light illuminating the slide. The higher the magnification, the greater the amount of light that is needed.

The light source is at the base of most microscopes. Usually, the light source is set at maximum and the amount of light on the slide is adjusted with the iris diaphragm.

Precautions for Proper Use and Care of the Microscope

1. Carry the microscope with both hands. Keep it upright. If the microscope is inverted, the eyepieces may fall out.

2. Do not remove the objective or ocular lenses for any purpose.

3. If something seems stuck or you have problems making adjustments, do not apply force. Consult the instructor.

4. Never touch or wipe the lenses with anything but lens paper. Clean the lens by gently drawing a flat piece of lens paper across it. The presence of foreign particles can be determined by rotating the eyepiece lenses manually as you

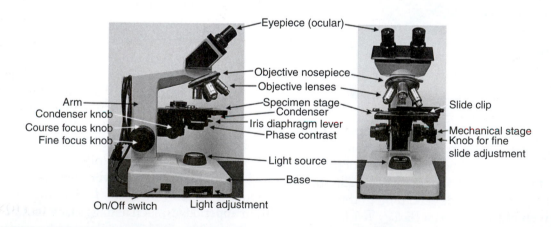

Figure 3.2 Modern bright-field compound microscope. Courtesy of Carl Zeiss, Inc.

look through the microscope. A pattern that rotates is evidence of dirt. If wiping the lenses with lens paper does not remove the dirt, consult the instructor. It may be on the inside surface of the lens.

5. Before storing the microscope:

    If the microscope has an adjustable tube, rack it down so that the microscope can be stored more easily.

    Make certain the eyepiece lenses are clean. Sweat deposits from your eyes can etch the glass.

    Important: If you have used the oil immersion lens, be sure to wipe off the oil. If not removed, it can leak into the lens and cause severe damage.

Definitions

Compound microscope. A microscope with more than one lens system.

Condenser. A structure located below the microscope stage that contains a lens for focusing light on the specimen as well as an iris diaphragm.

Iris diaphragm. An adjustable opening that regulates the amount of light illuminating the specimen.

Magnification. The microscope's ability to optically increase the specimen size.

Resolution. The smallest separation that two structural forms—two adjacent cilia, for example—must have in order to be distinguished optically as separate.

Wet mount. A laboratory technique in which a microscopic specimen in liquid is placed on the surface of a slide and covered with a coverslip.

Objectives

1. Describe the parts of the microscope.
2. Describe stained and unstained material.
3. Explain the use and proper care of the microscope.

Prelab Questions

1. Why should you carry a microscope in the upright position?
2. What is the only material that can be used to wipe a lens?

3. Why must you wipe off the eyepiece lens before storing the microscope?

Materials

Prepared slides of a printed letter e with coverslips (optional)

Prepared stained slides of protozoa and other large cells

Several jars of hay infusion containing protozoa and algae

Suspension of bread yeast

PROCEDURE

Important note: In this introduction to the microscope, the 100× oil objective lens (usually labeled with a broad band) will not be used. It is the most expensive lens and its use with immersion oil will be explained in exercise 4. Be particularly careful not to hit this lens on the stage.

1. Place the microscope on a clear space on your bench, away from any flame or heat source. Identify the different parts with the aid of figure 3.2 and the Getting Started section, "The Parts of the Microscope."

2. Before using the microscope, be sure to read the Getting Started section, "Precautions for Proper Use and Care of the Microscope."

3. Obtain either a prepared slide with the printed letter e covered with a coverslip or a large stained specimen of protozoa, fungi, or algae. Place the slide in the mechanical stage, coverslip side up, and turn on the light at the base of the microscope. Move the eyepiece lenses apart, and then while looking through the lenses, push them together until you see one circle. This is the ocular width of your eyes. If you note the number on the scale between the lenses, you can set the microscope to this number each time you use it.

4. Rotate the low-power objective lens (10×) into place. When looking from the side, bring the lens as close to the slide as possible

(or the slide to the lens depending on the microscope). Then when looking through the microscope, use the large knob, called the coarse-adjustment knob, to raise the lens (or lower the stage) until the object is in focus. Increase and decrease the amount of light with the iris diaphragm lever to determine the optimal amount of light. Continue to focus using the fine-adjustment knob, which is the smaller knob. Remember that the objective lens should never touch the surface of the slide or coverslip.

5. Move the slide back and forth. When viewing objects through the microscope, the image moves in the opposite direction than the slide is actually moving. It takes awhile to become accustomed to this phenomenon, but later it feels normal.

6. View the specimen at a higher magnification. Rotate the 40× objective lens into place (sometimes called the high–dry objective lens). Be very careful not to hit the slide or the stage with the oil immersion lens. Notice the change in the amount of light needed. Also note that the specimen is almost in focus. Most microscopes are parfocal, meaning that the objective lens can be rotated to another lens and the slide remains in focus. View several prepared slides until you become comfortable with the microscope. Always start focusing with the low-power objective lens, then move to the high–dry objective lens.

Wet Mounts

1. Prepare a wet mount (figure 3.3). Clean a glass slide with a mild cleansing powder such as Bon Ami or as directed by your instructor.

2. Place a drop (about the size of a small pea) of hay infusion or other material on the slide using a dropper. Carefully place a coverslip on edge next to the drop and slowly lower it so that it covers the drop. Try to avoid air bubbles. If the drop is so big it leaks out from under the coverslip, discard the slide into a container designated by the instructor and prepare another wet mount.

3. Examine the wet mount using the low-power objective lens, and then switch to the high-power objective lens. When viewing unstained material, it is necessary to reduce the amount of light to increase the contrast between the cells and the liquid. If you are having difficulty focusing on the material, try to focus on the edge of the coverslip and then move the slide into view.

4. Draw examples of the material viewed. Indicate the total magnification.

5. Dispose of your slides as directed by your instructor. The material on the slide is still viable, so the slides should be boiled or autoclaved.

6. Wipe the lens with lens paper. Return your microscope carefully to its storage space.

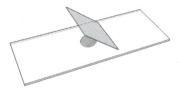

Figure 3.3 Preparation of a wet mount.

EXERCISE

3

Laboratory Report: Introduction to the Compound Light Microscope

Results

1. Draw four fields of the material you observed. Indicate which objective lens was used and the total magnification. If possible, show the same field or material at two different magnifications.

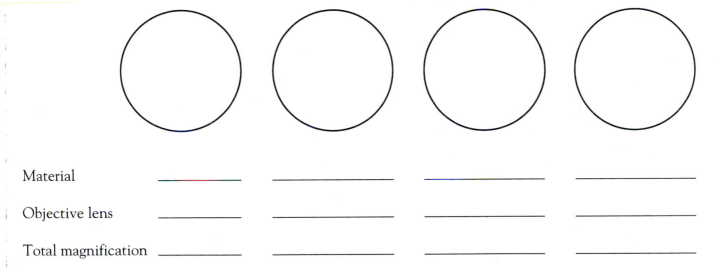

Material   _____   _____   _____   _____

Objective lens   _____   _____   _____   _____

Total magnification   _____   _____   _____   _____

2. On your microscope, what is the magnification of
   a. the eyepiece lenses?
   b. the low-power lens?
   c. the high-dry lens?

Questions

1. When you increase the magnification, is it necessary to increase or decrease the amount of light?

2. When looking at unstained material, do you need more or less light than what is needed to view a stained preparation?

3. Why is it convenient to have a parfocal microscope?

4. Why couldn't you see a virus with your microscope even if you increased the eyepiece lens magnification to 100×?

5. Why was Koch's observation of bacteria in blood so significant?

6. When observing a speciman through the microscope, how do you calculate the total magnification?

4 The Oil Immersion Lens

Getting Started

To observe bacteria, either stained or in a wet mount, it is necessary to use the oil immersion lens. The magnification of this lens is 100× and when viewing a specimen through an eyepiece of 10×, the total magnification is 1,000×, about the maximum possible with the light microscope. This lens is the longest and usually has a colored band around it. Extra care must be taken when using it because not only is it the most valuable, but because of its length, it is possible to hit and break the slide when attempting to focus with it. Also, care must be taken not to hit the lens on the stage when rotating a lens into place.

The reason this lens is called the oil immersion lens is that **immersion oil** is put on the top of a stained slide or wet mount and the lens is then carefully lowered into the oil (figure 4.1). The purpose of the oil is to prevent **refraction** of the light entering the lens. Light bends each time it goes through a different medium. (This can be seen in the appearance of a spoon in a glass of water.) To prevent the light's bending from its source in the microscope through air and then through the glass lens, the **refractive index** of the oil is the same as that of glass. This can be observed if your dispensing bottle for oil has

a glass dropper. The dropper will disappear when you immerse it in the oil because the light does not bend between the immersion oil and the glass (see Appendix 6 for additional information).

In this exercise, the oil immersion lens will be used to examine stained slides. Observing wet mounts to determine whether bacteria are motile is optional. This is a fairly difficult procedure and can be done later when you have had more experience using the microscope.

Definitions

Immersion oil. Oil placed on a slide to minimize refraction of the light entering the lens.
Refraction. The bending of light as it passes from one medium to another.
Refractive index. The ratio of the velocity of light in the first of two media to its velocity in the second medium as it passes from one medium into another medium.

Objectives

1. Explain the use of the oil immersion lens when examining stained slides.

2. Explain the use of the oil immersion lens to observe motile bacteria in a wet mount preparation (optional).

Prelab Questions

1. What is the total magnification of your microscope when you are observing stained bacteria with the oil immersion lens?

2. How can the refraction of light be reduced as it goes through the glass slide and into the lens?

3. When viewing stained slides, which lens do you use first to focus the microscope?

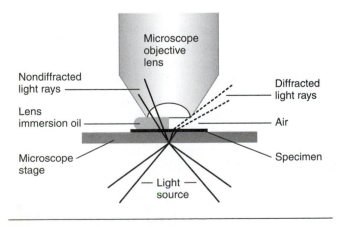

Figure 4.1 Oil immersion lens.

Materials

Prepared stained slides

Cultures of the following (for optional wet mounts):

Overnight cultures in nutrient broth of the following:

Spirillum volutans

Bacillus cereus

Pseudomonas

Overnight culture in trypticase soy broth of

Staphylococcus epidermidis

Vaseline (optional)

Toothpicks (optional)

PROCEDURE

The Oil Immersion Lens

1. Place a stained slide in the mechanical stage. Using the low-power 10× lens, first focus on the specimen with the coarse-adjustment knob, then with the fine-adjustment knob (see exercise 3). Increase or decrease the light with the iris diaphragm for optimal viewing. Repeat with the high–dry power.

2. Swing the lens away from the slide and place a drop of immersion oil directly on top of the slide.

3. Carefully rotate the oil immersion lens into the oil and click into place. Focus the microscope using the fine-adjustment knob. If the microscope is parfocal, it should take very little adjustment. Never move the oil immersion lens toward the slide while looking through the microscope, because you can hit the slide with the lens.

4. Notice that you must increase the amount of light as you increase the magnification.

5. Examine the stained slides to become familiar with the appearance of stained bacteria. This will be useful when you prepare your own slides.

6. Draw examples of several fields you observed. After you feel comfortable using the oil immersion lens, swing it away from the slide, remove the slide, and take the oil off the slide with a piece of lens paper or as directed by your instructor.

7. **Important:** Before storing the microscope, remove the oil from the lens by carefully wiping the lens with several flat pieces of lens paper. If the oil is not removed, it can seep around the cement holding the lens in place, damaging the microscope. Do not crumple the paper, but slide it over the lens. Also carefully wipe the eyepiece with lens paper.

Determining Motility with a Wet Mount (optional)

Some bacteria can move through liquid using flagella (singular, *flagellum*). These flagella are much smaller than those seen on protozoa and can only be visualized in the light microscope if special stains are used.

Although bacterial cell motility is usually determined by the semisolid agar stab inoculation method, it is sometimes determined by direct microscopic examination of wet mounts. This has the advantage of immediate results and also gives some additional information about the shape and size of the organism. Sometimes it is possible to see how the cells are arranged, as in tetrads or chains. However, these groupings are easily broken up in the process of making the wet mount so the designations are not reliable.

There are also disadvantages for determining motility from a wet mount. Some types of media inhibit motility, and cultures not actively multiplying may no longer be motile.

When observing bacteria in wet mounts, it is important to distinguish true motility from other kinds of motion. Evaporation at the edge of the coverslip causes convection currents to form. The suspended cells appear to be moving along in a stream flowing to the edge of the coverslip. If cells are truly motile, they swim in random directions.

"Brownian movement" is a form of motion caused by molecules in liquid striking a solid object—in this case, a bacterial cell—causing the cell to slightly bounce back and forth. Bacterial cells appear to jiggle in liquid, but if the cell is actually motile it moves from one point to another.

PROCEDURE

1. Using aseptic technique, sterilize a loop in the Bunsen burner, permit it to cool, and place a drop of culture on a clean slide. The drop should be about the size of a small pea. Remember to sterilize the loop before putting it down. It is easier to use a separate slide for each culture when beginning.

2. Carefully place a coverslip on one edge next to the drop and lower it down (see figure 3.3). If some of the drop seeps out from under the coverslip, it may contaminate the lens. Dispose of the slide as directed by the instructor—it cannot be simply washed off because it contains viable microorganisms.

3. Focus the slide as usual, first with the low-power lens, then with the high-dry lens. You will have to greatly reduce the amount of light using the iris diaphragm because there is very little contrast between the broth and the unstained organisms.

4. Place a drop of immersion oil on the coverslip of the wet mount and carefully click the oil immersion lens into place. Refocus the microscope, being very careful not to touch the slide with the lens. You should only need to use the fine-adjustment knob.

5. Observe the bacteria to determine if they are motile. Distinguish among Brownian movement, convection currents, and true motility (see the Getting Started section).

6. If you want to have more time to observe your wet mounts, you can coat the edge of the coverslip with vaseline using a toothpick before covering the drop on the slide. This will prevent the specimen from evaporating as quickly.

7. Dispose of your slides as directed by your instructor.

8. Record your results.

EXERCISE

4 Laboratory Report: The Oil Immersion Lens

Results

Make drawings of the slides as observed under the oil immersion lens.

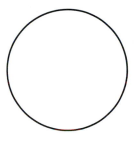 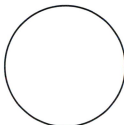 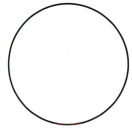 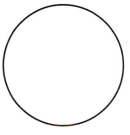

Questions

1. How do you adjust the amount of light when viewing a slide through the microscope?

2. Why is it important to remember that the 100× lens is longer than the lower-power lens?

3. When increasing the magnification while observing a stained slide, do you increase or decrease the amount of light?

4. What **must** you do before storing your microscope after using the oil immersion lens?

Wet Mounts (Optional)

5. How can you distinguish true motility from other movement?

6. How do cells appear when Brownian movement is causing their motion?

7. How do cells appear when convection currents are causing their motion?

INTRODUCTION to Staining of Microorganisms

Bacteria are difficult to observe in a broth or wet mount because there is very little contrast between them and the liquid in which they are suspended. This problem is solved by staining bacteria with dyes. Although staining kills bacteria so their motility cannot be observed, the stained organisms contrast with the surrounding background and are much easier to see. The determination of the shape, size, and arrangement of the cells after dividing is useful in identifying an organism. These can be demonstrated best by preparing a smear on a glass slide from clinical material, a broth culture, or a colony from a plate, then staining the smear with a suitable dye. Examining a stained preparation is one of the first steps in identifying an organism.

Staining procedures used here can be classified into two types: the simple stain and the multiple stain. In the **simple stain,** a single stain such as methylene blue or crystal violet is used to dye the bacteria. The most commonly used simple stains are basic dyes. These dyes are positively charged, making it possible to stain the negatively charged bacteria.

The shape and the grouping of the organisms can be determined, but all organisms (for the most part) are stained the same color.

Another kind of simple stain is the negative stain. In this procedure, the organisms are mixed with a dye on a slide and the mixture is permitted to air dry. When the stained slide is viewed under the microscope, the organisms are clear against a dark background.

The multiple stain involves more than one dye. The best known example is the differential Gram stain, which is widely used. After staining, some organisms appear purple and others pink, depending on the structure of their cell wall.

Special stains are used to observe specific structures of bacteria. Compared with eukaryotic organisms, prokaryotic organisms have relatively few morphological differences. Several of these structures, such as endospores, capsules, acid-fast cell walls, storage bodies, and flagella, can be seen with special stains. In the next two exercises, you will have an opportunity to stain bacteria with a variety of staining procedures and observe these structures.

Notes:

Notes:

EXERCISE

5

Simple Stains: Positive and Negative Stains

Getting Started

Two kinds of simple stains will be done in this exercise: the simple stain and the negative stain. Microbiologists most frequently stain organisms with the Gram stain, but in this exercise a simple stain will be used to give you practice staining and observing bacteria before doing the more complicated multiple, or **differential, stains.**

After you have stained your bacterial **smears,** you can examine them with the oil immersion lens, which will allow you to distinguish the morphology of different organisms. The typical bacteria you will see are about 0.5–1.0 **micrometers** (µm) in width to about 2–7 µm long and are usually rods, cocci, or spiral-shaped. Sometimes rods are referred to as *bacilli,* but since that term is also a genus name (*Bacillus*) for a particular organism, the term *rod* is preferred.

Another kind of simple stain is the **negative stain** (color plate 4). Although it is not used very often, it is advantageous in some situations. Organisms are mixed in a drop of nigrosin or India ink on a glass slide. After drying, the organisms can be observed under the microscope as clear areas in a black background. This technique is sometimes used to observe **capsules** or **inclusion bodies.** It also prevents eyestrain when many fields must be scanned. The dye tends to shrink away from the organisms, causing cells to appear larger than they really are.

In both of these simple stains, you will be able to determine the shape of the bacteria and the characteristic grouping after cell division (as you did in the wet mounts). Some organisms tend to stick together after dividing and form chains or irregular clumps. Others are usually observed as individual cells. However, this particular characteristic depends somewhat on how the organisms are grown. *Streptococcus* bacteria form long, fragile chains in broth, but if they grow in a colony on a plate, it is sometimes difficult to make a smear with these chains intact.

Definitions

Capsule. A gelatinous material coating the outside of the cell.

Differential stain. A procedure that stains specific morphological structures—usually a multiple stain.

Inclusion bodies. Granules of storage material such as sulfur that accumulate within some bacterial cells.

Micrometer (abbreviated µm). The metric unit used to measure bacteria. It is 10^{-6} m (meter) and 10^{-3} mm (millimeter). One meter = 1,000 mm.

Negative stain. A simple stain in which the organisms appear clear against a dark background.

Parfocal. All the objective lenses of the microscope are in the same plane, so that it is not necessary to refocus when changing lenses.

Simple stain. A procedure for staining bacteria consisting of a single stain.

Smear. A dried mixture of bacteria and water (or broth) placed on a glass slide in preparation for staining.

Objectives

1. Describe the preparation and staining of a bacterial smear using a simple stain.
2. Describe the preparation of a negative stain.
3. Describe the morphologies and arrangements of bacteria observed in a simple stain.

Prelab Questions

1. What is the dried mixture of cells and water on a slide called?
2. How big is the average bacterial cell?
3. Name two kinds of information you can learn from a simple stain.

Materials

Cultures of the following:
  Bacillus subtilis or Bacillus cereus
  Staphylococcus epidermidis
  Streptococcus mutans
  Micrococcus luteus
Staining bottles with
  crystal violet
  methylene blue
  safranin
Glass slides
Waterproof marking pen or wax marking pen
Tap water in small dropper bottle (optional)
Inoculating loop
Alcohol sand bottle (a small screw cap bottle half full of sand and about three-quarters full of 95% alcohol) (optional) (figure 5.1)
Disposable gloves (optional)

PROCEDURE

Simple Stain

1. Clean a glass slide by rubbing it with slightly moistened cleansing powder such as Boraxo or Bon Ami. Rinse well and dry with a paper towel. New slides should be washed because sometimes they are covered with a protective coating.

2. Draw one or two circles with a waterproof pen or wax pencil on the underside of the slide.

Figure 5.1 Alcohol sand bottle and inoculating loop.

If the slide has a frosted portion, you can also write on it with a pencil. This is useful because it is easy to forget the order in which you placed the organisms on the slide and you can list them, for instance, from left to right (figure 5.2).

3. Add a drop of water to the slide on top of the circles. Use your loop to transfer tap water or use water from a dropper bottle. This water does not need to be sterile. Although there are some organisms in municipal water systems, they are nonpathogens and are too few to be seen.

If you are preparing a smear from a broth culture, as you will do in the future, add only the broth to the slide. Broth cultures are relatively dilute, so no additional water is added.

4. Sterilize a loop by holding it at an angle in the flame of the Bunsen burner. Heat the entire wire red hot, but avoid putting your hand directly over the flame or heating the handle itself (figure 5.3).

5. Allow the loop to cool a few seconds, then remove a small amount of a bacterial culture and suspend it in one of the drops of water on the slide (see figure 5.3). Continue to mix in bacteria until the drop becomes slightly cloudy. If your preparation is too thick, it will stain unevenly, and if it is too thin, you will have a difficult time finding organisms under the microscope. In the beginning, it may be better to err on the side of having a slightly too cloudy preparation—at least you will be able to see organisms and you will learn from experience how dense to make the suspension.

6. Heat the loop red hot again. It is important to burn off the remaining organisms so that you will not contaminate your benchtop. If you then rest your loop on the side of your Bunsen burner, it can cool without burning anything on the bench.

Figure 5.2 Slide with two drops of water. Two different bacteria can be stained on one slide.

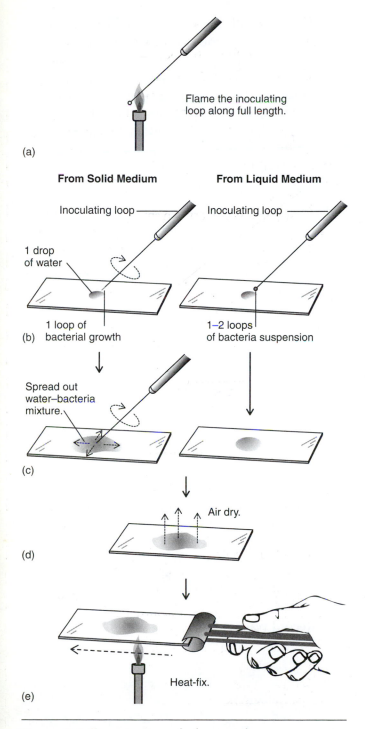

(a) Flame the inoculating loop along full length.

From Solid Medium

Inoculating loop

1 drop of water

(b) 1 loop of bacterial growth

From Liquid Medium

Inoculating loop

1–2 loops of bacteria suspension

(c) Spread out water–bacteria mixture.

(d) Air dry.

(e) Heat-fix.

Figure 5.3 Preparation of a bacterial smear.

Sometimes the cell material remaining on the loop spatters when heated. To prevent this, some laboratories remove bacterial cell material from the loop by dipping the loop in an alcohol sand bottle. Then the loop is heated red hot in the Bunsen burner.

7. Permit the slide to dry. Do not heat it in any way to hasten the process, because the cells will become distorted. Place the slide off to the side of the bench so that you can proceed with other work.

8. When the slide is dry (in about 5–10 minutes), heat-fix the organisms to the slide by quickly passing it through a Bunsen burner flame two or three times so that the bottom of the slide is barely warm. This step causes the cells to adhere to the glass so they will not wash off in the staining process (figure 5.3).

9. Place the slide on a staining loop over a sink or pan. Alternatively, hold the slide over the sink with a forceps or clothespin. Cover the specimen with a stain of your choice—crystal violet is probably the easiest to see (figure 5.4).

10. After about 20 seconds, pour off the stain and rinse with tap water (figure 5.4).

11. Carefully blot the smear dry with a paper towel or bibulous paper. Do not rub the slide from side to side as this will remove the organisms. Permit the slide to air dry until you are sure the slide is completely dry (figure 5.4).

12. Observe the slide under the microscope. Because you are looking at bacteria, you must use the oil immersion lens in order to see them. Focus the slide on low power and, then again on high power. Cover the smear with immersion oil and move the immersion lens into place. If your microscope is **parfocal,** it should be very close to being in focus. Note that no coverslip is used when looking at stained organisms.

Remember, never move the immersion lens toward the slide while looking through the microscope. You may hit the slide with the lens and damage the lens. When you have a particularly thin smear, it is sometimes helpful to put a mark on the slide near the stain using a marking pen. It is easy to focus on the pen mark, and you will know that you have the top of the slide in focus and can then search for the smear.

13. Record your results.

14. You can save your stained slide with the oil on it. If you do not want to save the slide, simply clean it with cleanser and water. The staining

Simple Staining Procedure

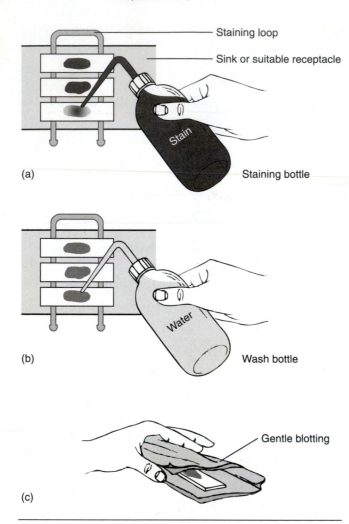

(a)

Staining loop

Sink or suitable receptacle

Stain

Staining bottle

(b)

Water

Wash bottle

(c)

Gentle blotting

Figure 5.4 (*a*) Staining, (*b*) washing, and (*c*) blotting a simple stain. From John P. Harley and Lansing M. Prescott, *Laboratory Exercises in Microbiology*, 5th ed. copyright © 2002 The McGraw-Hill Companies. All Rights Reserved. Reprinted by permission.

procedure kills the bacteria, so the slide does not need to be boiled before cleaning.

15. **Important:** It is essential to wipe off the oil from the immersion lens with a flat piece of lens paper before storing the microscope. Do not crumple the paper, as this could harm the lens. Also be sure to wipe the eyepiece with lens paper.

Negative Stain

The negative stain can be used to observe capsules or inclusion bodies. However, in this exercise, the negative stain will be used to compare the appearance of the same organisms using the two staining procedures.

Materials

Same cultures used for simple stain

Bottle of India ink or nigrosin

PROCEDURE

1. Place a drop of water on a clean slide and add organisms with a sterile loop until the drop is slightly cloudy.

2. Sterilize the loop, mix a loopful of India ink into the drop, and spread the mixture out into a thin film.

3. Sterilize the loop.

4. Let dry and examine under the microscope. Bacteria can be seen as clear areas on a black background.

5. Record your results.

EXERCISE

Laboratory Report: Simple Stains:
Positive and Negative Stains

Results

1. Simple stain

	Staphylococcus *epidermis*	Bacillus *subtilis*	Micrococcus *iuhus*	Streptococcus *p phnuemonia*
Draw shape and arrangement.	cocci {round} – packets of 4	long. rod shape – single s	cocci – tetrads	cocci – chains diplococci

Gram.

2. ~~Negative~~ stain

           +   +   +

Questions

1. What three basic shapes of bacteria can be seen in a simple stain?

    coccus – round
    bacillus – rod
    vibro – curvyrod
    spirochete. – spiral
    pnelomorphie – changeeble shaped b/c no vigia

2. How does the appearance of the negative stain compare to the appearance of the simple stain?

3. Why do you gently heat the slide before staining?

to adhere the cells / microorganisms to the slide

4. What might happen if you heat-fix a slide before it is dry?

you could lose sample w/ evaporation

5. When blotting dry a stained slide, what will happen if you rub it from side to side?

you'll wipe the stain or organisms off

10^m

micro meters.

6. How many μm are in a millimeter (mm)? *1000.*
   How many μm are in a meter (m)? *1 million. 1,000,000*

7. (To ponder) If a large dill pickle is 100 millimeters long (or 10 centimeters) and you magnified it 1,000×, how long would it appear? (Assume a yard is approximately a meter.)
   a. as long as a table tennis table (~3 meters)
   b. as long as a swimming pool (25 meters)
   c. as long as a football field (100 meters)

Hint: 10 centimeters is 0.1 meter.

EXERCISE

6

Differential and Other Special Stains

A Differential Stain: The Gram Stain

Getting Started

Differential stains usually involve at least two dyes and are used to distinguish one group of organisms from another. For example, the Gram stain determines whether organisms are Gram positive or Gram negative.

The Gram stain is especially useful as one of the first procedures employed in identifying organisms. It reveals not only the morphology and the arrangement of the cells but also information about the cell wall.

In the late 1800s, Christian Gram devised the staining procedure when trying to stain bacteria so that they contrasted with the tissue sections he was observing. Many years later, it was found that purple (Gram-positive) bacteria had thick cell walls of **peptidoglycan,** while pink (Gram-negative) bacteria had much thinner cell walls of peptidoglycan surrounded by an additional lipid membrane. The thick cell wall retains the purple dye in the procedure, but the thin wall does not (table 6.1).

In the Gram stain, a bacterial smear is dried and then heat-fixed to cause it to adhere to the glass slide (as in the simple stain). It is then stained with crystal violet dye, the primary stain, which is rinsed off and replaced with an iodine solution. The iodine acts as a **mordant**—that is, it binds the dye to the cell. The smear is then decolorized with alcohol and **counterstained** with safranin. In Gram-positive organisms, the purple crystal violet dye, complexed with the iodine solution, is not removed by the alcohol, and thus the organisms remain purple. On the other hand, the purple stain is removed from Gram-negative

organisms by the alcohol, and the colorless cells take up the red color of the safranin counterstain.

Note: In the past, many clinical laboratories used a 50/50 mixture of alcohol and acetone because it destains faster than 95% alcohol. However, most labs now use 95% alcohol, which is just as effective, but the stain must be decolorized longer (up to 30 seconds).

Special Notes to Improve Your Gram Stains

1. Gram-positive organisms can lose their ability to retain the crystal violet complex when they are old. This can happen when a culture has been incubating 18 hours or more. The genus *Bacillus* is especially apt to become Gram negative. Use young, overnight cultures whenever possible. It is interesting to note that Gram-positive organisms can appear Gram negative, but Gram-negative organisms almost never appear Gram positive.

2. Another way Gram-positive organisms may appear falsely Gram negative is by over decolorizing in the Gram-stain procedure. If excessive amounts of alcohol are used, almost any Gram-positive organism will lose the crystal violet stain and appear Gram negative.

3. If you are staining a very thick smear, it may be difficult for the dyes to penetrate properly. This is not a problem with broth cultures, which are naturally quite dilute, but be careful not to make the suspension from a colony in a drop of water too thick.

4. When possible, avoid making smears from inhibitory media such as eosin methylene blue (EMB) because the bacteria frequently give variable staining results and can show atypical morphology.

5. The use of safranin in the Gram stain is not essential. It is simply used as a way of dying the colorless cells so they contrast with the purple. If you are color-blind and have difficulty distinguishing pink from purple, try other dyes as a counterstain.

Table 6.1 Appearance of the Cells After Each Procedure

	Gram +	Gram –
Crystal violet	Purple	Purple
Iodine	Purple	Purple
Alcohol	Purple	Colorless
Safranin	Purple	Pink

Definitions

Counterstain. A stain used to dye unstained cells a contrasting color in a differential stain.

Mordant. A substance that increases the adherence of a dye.

Peptidoglycan. The macromolecule making up the cell wall of most bacteria.

Vegetative cell. A cell that has not formed spores or other resting stages.

Objectives

1. Describe the Gram-stain procedure.
2. Differentiate between Gram-positive organisms and Gram-negative organisms.
3. Describe other special stains.

Prelab Questions

1. What structure of the bacterial cell determines whether the organism is Gram positive or Gram negative?
2. What is the composition of both Gram-positive and Gram-negative cell walls?
3. What color are Gram-positive cells after Gram staining?

Materials

Staining bottles of the following:

crystal violet

iodine

95% alcohol

safranin

Clothespin or forceps

Staining bars

Disposable gloves

Overnight cultures growing on trypticase soy agar slants of the following:

Escherichia coli

Bacillus subtilis

Staphylococcus epidermidis

Streptococcus mutans

Micrococcus luteus

PROCEDURE

Procedure for Gram Stain

1. Put two drops of water on a clean slide. In the first drop, make a suspension of the unknown organism to be stained just as you did for a simple stain (see exercise 5 for instructions on preparing a smear). In the second drop, mix together a known Gram-positive organism and a known Gram-negative organism. This mixture is a control to ensure that your Gram-stain procedure (figure 6.1) gives the proper results. Heat-fix the slide after the slide is dry.

2. Place the slide on a staining bar across a sink (or can). Alternatively, hold the slide with a clothespin or forceps over a sink.

3. Flood the slide with crystal violet until the slide is completely covered. Leave the stain on for 6–30 seconds and then discard into the sink. The timing is not critical. Rinse the slide with water from a wash bottle or with gently running tap water (figure 6.1, *a* and *b*).

4. Flood the slide with iodine for about 12–60 seconds and then wash with water (*c* and *d*).

5. Hold the slide at a 45° angle and carefully drip 95% ethanol, the decolorizer, over it until no more purple dye runs off. Immediately wash the slide with tap water. Thicker smears may take longer than thinner ones, but ethanol should usually be added for about 20 seconds. Timing is critical in this step (*e* and *f*).

6. Flood the slide with safranin and leave it on for 10–30 seconds—timing is not important. Wash with tap water. Safranin is a counterstain because it stains the cells that have lost the purple dye (*g* and *h*).

7. Blot the slide carefully with bibulous paper or paper towel to remove the water, but do not rub from side to side. When it is completely dry, observe the slide under the microscope. Remember that you must use the oil immersion lens to observe bacteria. Compare your stain to the control mixture on the same slide and with color plate 5.

8. Describe the appearance of your stained bacteria in the Results section of the Laboratory Report.

9. Be sure to carefully remove the immersion oil from the lens with lens paper before storing the microscope.

(a) Apply crystal violet to smear; 6–30 seconds.

(b) Rinse with water for 5 seconds.

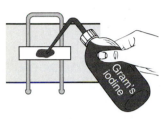

(c) Cover with Gram's iodine; 12–60 seconds.

(d) Rinse with water for 5 seconds.

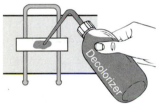

(e) Decolorize 1–5 seconds (acetone–alcohol). 5–30 seconds (alcohol).

(f) Rinse with water for 5 seconds.

(g) Counterstain with safranin; 10–30 seconds.

(h) Rinse with water for 5 seconds.

(i) Blot dry with bibulous paper or a paper towel.

Figure 6.1 Gram-stain procedure.

A Differential Stain: Acid-Fast Stain (Optional)

Getting Started

The acid-fast stain is useful for identifying bacteria with a waxy lipid cell wall. Most of these organisms are members of a group of bacteria called mycobacteria. Although there are many harmless bacteria in this group, it also includes *Mycobacterium tuberculosis*, which is the cause of tuberculosis in humans. These organisms have a Gram-positive cell wall structure, but the lipid in the cell wall prevents staining with the Gram-stain dyes.

In the Ziehl–Neelsen (Kinyoun modification) acid-fast stain procedure, the dye carbolfuchsin stains the waxy cell wall. Once the lipid-covered cell has been dyed, it cannot be decolorized easily—even with alcohol containing HCL (called acid–alcohol). Nonmycobacteria are also dyed with the carbolfuchsin but are decolorized by acid–alcohol. These colorless organisms are stained with methylene blue, so they contrast with the pink acid-fast bacteria that were not decolorized.

The reason this stain is important is that one of the initial ways tuberculosis is diagnosed is by the presence of *Mycobacterium* in a patient's sputum. (Sputum is a substance that is coughed up from the lungs and contains puslike material.) Tuberculosis is a very serious disease worldwide and is now seen in the United States after decreasing for about 80 years. Because the process of finding acid-fast organisms in sputum is quite difficult and time-consuming, this test is usually performed in state health laboratories.

Objectives

1. Interpret acid-fast stain.
2. Examine acid-fast organisms.

Materials

Culture of the following:
   Mycobacterium smegmatis mixed with *Staphylococcus epidermidis*
Carbolfuchsin in staining bottles
Methylene blue in staining bottles
Acid–alcohol in staining bottles
Beaker
Metal or glass staining bars

Procedure for Acid-Fast Stain (Kinyoun Modification)

1. Prepare a smear of the material and heat-fix (see exercise 5).
2. Cover the smear with carbolfuchsin and stain for 3–5 minutes. Do not heat (figure 6.2).
3. Rinse with water.
4. Decolorize with acid–alcohol for 10–30 seconds.
5. Rinse with water.

6. Counterstain with methylene blue for 20–30 seconds.
7. Rinse with water.
8. Blot dry carefully and examine under the oil immersion lens.
9. The acid-fast *Mycobacterium* will appear pink and the *Staphylococcus* will appear blue (color plate 6).
10. Record results.

(a) Apply carbolfuchsin to smear for 5 minutes.

(b) Rinse with water.

(c) Decolorize with acid–alcohol; 10–30 seconds.

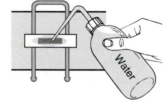

(d) Rinse with water.

(e) Counterstain with methylene blue; 20–30 seconds.

(f) Rinse with water.

(g) Blot dry with bibulous paper or paper towel.

Figure 6.2 Acid-fast stain procedure.

Special Stains: Capsule, Endospore, Storage Granules, and Flagella

Getting Started

Although bacteria have few cell structures observable by light microscopy when compared to other organisms, some have capsules, endospores, storage granules, or flagella. The following procedures will enable you to see these structures.

Objectives

1. Examine various structures and storage products of bacteria.
2. Interpret staining procedures for these structures.

Capsule Stain (Optional)

A capsule is a somewhat gelatinous coating surrounding the cell. It can consist of amino acids or carbohydrates and it can protect the bacterium from engulfment by white blood cells. The ability to produce a capsule frequently depends on the availability of certain sugars. *Streptococcus mutans*, for example, produces a capsule when growing on sucrose but not when growing on glucose.

Materials

Culture of the following:

Klebsiella or other organism with a capsule growing on a slant

India ink

Procedure for Capsule Stain

1. Make a suspension of the organism in a drop of water on a clean slide.
2. Put a drop of India ink next to it.
3. Carefully lower a coverslip over the two drops so that they mix together. There should be a gradient in the concentration of the ink.
4. Examine the slide under the microscope and find a field where you can see the cells surrounded by a halo in a black background. The halo is the capsule surrounding the cell (color plate 7).
5. After viewing, drop slides in a beaker or can of boiling water and boil for a few minutes before cleaning. This is necessary because the bacteria are not killed in the staining process.
6. Record results.

Endospore Stain (Optional)

Some organisms such as *Bacillus* and *Clostridium* can form a resting stage called an *endospore* (or sometimes just *spore*), which protects them from heat, chemicals, and starvation. When the cell determines that conditions are becoming unfavorable due to a lack of nutrients or moisture, it forms an endospore. When conditions become favorable again, the spore can germinate and the cell can continue to divide. The endospore is resistant to most stains so special staining procedures are needed.

> ### Materials
>
> Culture of the following:
>
> *Bacillus cereus* on a nutrient agar slant after 3–4 days' incubation at 30°C
>
> Malachite green in staining bottles
>
> Safranin in staining bottles
>
> Metal or glass staining bars
>
> Beaker or can

Procedure for Endospore Stain

1. Prepare a smear on a clean slide and heat-fix.
2. Add about an inch of water to a beaker and bring it to a boil.

3. Place two short staining bars over the beaker and place the slide on them.
4. Tear a piece of paper towel a little smaller than the slide and lay on top of the smear. The paper prevents the dye from running off the slide.
5. Flood the slide with malachite green and steam for 5 minutes. Continue to add stain to prevent the dye from drying on the slide (figure 6.3).
6. Decolorize with water for about 30 seconds by flooding with water or holding under gently running tap water. The **vegetative cells** (dividing cells) lose the dye, but the endospores retain the dye.
7. Counterstain with safranin for about 30 seconds and then wash with tap water for 30 seconds. Blot dry carefully.
8. Observe with the oil immersion lens. The endospores appear green and the vegetative cells appear pink (color plate 8). Sometimes

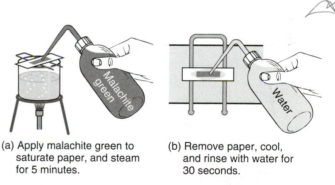

(a) Apply malachite green to saturate paper, and steam for 5 minutes.

(b) Remove paper, cool, and rinse with water for 30 seconds.

(c) Counterstain with safranin for 30 seconds.

(d) Rinse with water for 30 seconds.

(e) Blot dry with bibulous paper or paper towel.

Figure 6.3 Procedure for endospore stain.

Spore Stain of *Bacillus* with Malachite Green

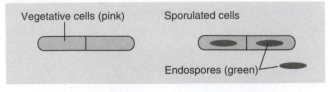

Vegetative cells (pink)   Sporulated cells

Endospores (green)

Gram Stain of *Bacillus*

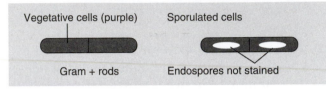

Vegetative cells (purple)   Sporulated cells

Gram + rods   Endospores not stained

Figure 6.4 Appearance of endospores stained with spore stain and Gram stain. *Note:* The *Bacillus* bacteria frequently lose their ability to stain Gram positive.

the endospore is still seen within the cell, and its shape and appearance can be helpful in identifying the organism. In other cultures, the endospores may be free because the cells around them have disintegrated (figure 6.4).

9. Record results.

10. Prepare and observe a Gram stain of the same culture (optional).

Note: When bacteria containing endospores are Gram stained, the endospores do not stain and the cells appear to have holes in them (figure 6.4).

Storage Granules Stain (Optional)

Many organisms can accumulate material as storage granules that are abundant in their environment for use in the future. For example, phosphate can be stored as metachromatic granules (also called volutin granules). When organisms containing these granules are stained with methylene blue, the phosphate granules are stained a darker reddish-blue.

Materials

Culture of the following:
Spirillum grown in nutrient broth
Methylene blue in staining bottles

Procedure for Storage Granules Stain

1. Prepare a smear from the broth. It might be helpful to remove the organisms from the bottom of the tube with a capillary pipet. Place a drop on a clean slide. Dry and heat-fix.

2. Flood the slide with methylene blue for about 20–30 seconds.

3. Wash with tap water and blot dry.

4. Observe with the oil immersion lens. The metachromatic granules should appear as dark reddish-blue bodies within the cells.

5. Record results.

Flagella Stain (Optional)

Some bacteria have flagella (singular, *flagellum*) for motility. Their width is below the resolving power of the microscope so they cannot be seen in a light microscope (the flagella seen at each end of *Spirillum* in a wet mount is actually a tuft of flagella). Flagella can be visualized if they are dyed with a special stain that precipitates on them, making them appear much thicker. The arrangement of the flagella on bacteria is usually characteristic of the organism and can aid in identification. There are three basic kinds of arrangement of flagella: **peritrichous, polar,** and **polar tuft.**

Definitions

Peritrichous. The surface of the organism is covered with flagella, such as *E. coli.*

Polar. A single flagellum at one or both ends of the organism, such as *Pseudomonas.*

Polar tuft (lophotrichous). A tuft of flagella at one or each end of an organism, such as *Spirillum.*

Materials

Stained demonstration slides of *Escherichia coli* and *Pseudomonas*

Procedure for Flagella Stain

1. Observe slides with stained flagella of several organisms and note the pattern of flagella. It is difficult to perform this staining procedure, so prestained slides are recommended.

2. Record results.

EXERCISE

6

Laboratory Report: Differential and Other Special Stains

Results

	Gram Reaction	Arrangement (Sketch)
E. coli		
B. subtilis		
S. epidermidis		
Streptococcus mutans		
M. luteus		

Optional Stains	Organism	Appearance
Acid-fast		
Capsule		
Endospore		
Storage granules		
Flagella		

Questions

1. What is the function of each one of the Gram-stain reagents?

   a. crystal violet. stains all organisms purple.

   MASE

   b. iodine - mordanate, causes crystal violet to sher to cells
      binds the dye

   rinse

   c. ethanol - decolorizer, removes crystal violet from
      gram negative.

   rinse

   d. saframa - restains whatever was renen off in rinse.

2. In which step of the Gram stain is the timing critical?

   The ethanol b/c could take all color off

3. Give two reasons Gram-positive organisms sometimes appear Gram negative.

— technical error, decolorized too long.

4. What is the purpose of using a control in the Gram stain?

to increase reliability of results and have a test to compare it to

5. What is a capsule?

6. What are storage granules, and why are they important to the cell?

7. How does an endospore appear (draw and indicate color)

   a. when Gram stained?

   b. when spore stained?

8. Why can't you Gram stain an acid-fast organism?

9. Why do you need a special staining procedure for flagella?

INTRODUCTION to Microbial Growth

The ability to grow bacteria is very important for studying and identifying them. Organisms in the laboratory are frequently grown either in a broth culture or on a solid agar medium. A broth culture is useful for growing large numbers of organisms. Agar medium is used in a petri dish when a large surface area is important, as in a streak plate. On the other hand, agar medium in tubes (called slants) is useful for storage because the small surface area is not as easily contaminated and the tubes do not dry out as fast as plates. You will be able to practice using media in all these forms (figure I.7.1).

You will also use different kinds of media in this section. Most media are formulated so that they will support the maximum growth of various organisms, but other media have been designed to permit the growth of desired organisms and inhibit others (selective). Still other media have been formulated to change color or in some other way distinguish one bacterial colony from another (differential). These media can be very useful when trying to identify an organism.

It is also important to know how to count bacteria. You will have an opportunity to learn about several techniques and their advantages and disadvantages.

In these exercises no pathogenic organisms are used, but it is very important to treat these cultures as if they were harmful because then you will be prepared to work safely with actual pathogens. Also, almost any organism can cause disease if there are large numbers in the wrong place.

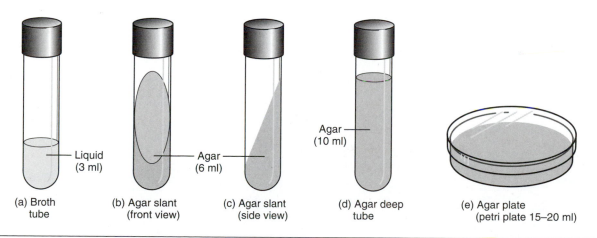

| (a) Broth tube | (b) Agar slant (front view) | (c) Agar slant (side view) | (d) Agar deep tube | (e) Agar plate (petri plate 15–20 ml) |

Liquid (3 ml) — Agar (6 ml) — Agar (10 ml)

Figure I. 7.1 Various media in different forms.

Notes:

INTRODUCTION to the Environment and Microbial Growth

An organism cannot grow and divide unless it is in a favorable environment. Environmental factors include temperature, availability of nutrients, moisture, oxygen, salinity, osmotic pressure, and presence of toxic products.

Each bacterial species has its own particular set of optimal conditions that allows maximum growth. These conditions probably reflect the environment in which the organism grows and competes with other organisms. For example, organisms found in a hot spring in Yellowstone National Park require a much higher temperature for growth in the laboratory than organisms isolated from the throat. In the next two exercises, we examine the effects of temperature and atmosphere (oxygen) on the growth of bacteria. (Osmotic pressure is examined in exercise 13.)

Notes:

Shake tube

insert rod
in (dont shake)

slant

just smeer on surface

10

The Effect of Incubation Temperature on Generation Time

Getting Started

Every bacterial species has an optimal temperature—the particular temperature resulting in the fastest growth. Normally, the optimal temperature for each organism reflects the temperature of its environment. Organisms associated with animals usually grow fastest at about 37°C, the average body temperature of most warm-blooded animals. Organisms can divide more slowly at temperatures below their optimum, but there is a minimum temperature below which no growth occurs. Bacteria usually are inhibited at temperatures not much higher than their optimum temperature.

The effect of temperature can be carefully measured by determining the **generation time** at different temperatures. Generation time, or doubling time, is the time it takes for one cell to divide into two cells; on a larger scale, it is the time required for the population of cells to double. The shorter the generation time, the faster the growth rate.

Generation time can only be measured when the cells are dividing at a constant rate. To understand when this occurs, it is necessary to study the growth curve of organisms inoculated into a fresh broth medium. If plate counts are made of the growing culture, it can be seen that the culture proceeds through the four phases of growth: lag, log, stationary, and death (figure 10.1).

In the lag phase, the cells synthesize the necessary enzymes and other cellular components needed for growth. The cells then grow as rapidly as the conditions permit in the log phase, and when there are no longer sufficient nutrients or toxic products accumulate, the cells go into the stationary phase. In the stationary phase, the number of viable cells neither increases nor decreases. This is followed by the death phase in which the cells die at a steady rate. The cells are growing at a constant maximum rate for the particular environment only in the log phase.

In this exercise, the generation time of *Escherichia coli* will be compared when growing at two different

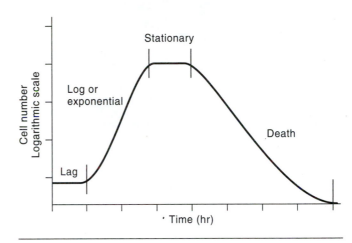

Figure 10.1 Growth curve showing the four phases of growth.

temperatures. The growth of the cells can be measured with a spectrophotometer or a Klett **colorimeter** because the number of cells in the culture is directly proportional to the absorbance (figure 10.2). That means that the **absorbance** (also called **optical density**) increases proportionately as the culture becomes increasingly more **turbid** (cloudy) from the multiplication of the bacteria. Readings of the incubating cultures are taken every 20 minutes for 80 minutes. The results are then plotted, and the generation time is determined.

Definitions

Absorbance. A measure of turbidity.

Colorimeter. An instrument used to measure the turbidity of bacterial growth.

Generation time. The time it takes for one cell to divide into two cells or for a population of cells to double.

Optical density (O.D.). An older but still widely used term for absorbance when used for measuring bacterial growth.

Turbid. A dense solution, cloudy.

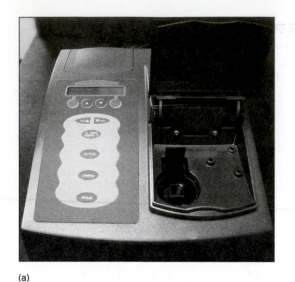

(a)

(b)

Figure 10.2 (*a*) Spectrophotometer. (*b*) Klett colorimeter. (*a*) © Anna Oller, University of Central Missouri. (*b*) Courtesy of VWR Scientific Company.

Objectives

1. Describe the phases of a growth curve.
2. Interpret the effect of temperature on generation time.
3. Describe the calculation of generation time.
4. Explain the use of semi-log paper.

Prelab Questions

1. What are the four phases of a bacterial growth curve?
2. What is the generation time of a bacterial cell?
3. Why is generation time only measured in the log phase?

Materials

Per team

Culture of the following:

  Escherichia coli (TS broth cultures in log phase)

Trypticase soy broth, in a tube that can be read in a spectrophotometer or Klett colorimeter and that has been prewarmed in a water bath, 1

Water bath at 30°C (to be used by half the class)

Water bath at 37°C (to be used by the other half of the class)

1ml pipet

PROCEDURE

1. Add 0.5 ml–1.0 ml of an *E. coli* culture growing in log phase to 5.0 ml trypticase soy broth. The dilution is not important as long as the broth is turbid enough to be read at the low end of the scale (0.1 O.D. or a Klett reading of about 50). If you start with an O.D. that is too high, your last readings will reach the part of the scale that is not accurate (an O.D. of about 0.4 or about 200 on the Klett).

2. With a wavelength of 420, set the spectrophotometer at zero with an uninoculated tube of trypticase soy broth (termed a *blank*). Your instructor will give specific directions.

3. Take a reading of the culture and record it as 0 time. Wipe off water and fingerprints from the tubes before taking a reading. Return the

tube to the assigned water bath as quickly as possible because cooling slows the growth of the organisms.

4. Read the O.D. of the culture about every 20 minutes for about 80 minutes. Record the exact time of the reading so the data can be plotted correctly.

5. Record your data—the time and O.D. readings—in your manual and on the blackboard.

6. Plot the data on semi-log graph paper (page 71). Semi-log paper is designed to convert numbers or data to $\log_{10}$ as they are plotted on the y axis. The same results would be obtained by plotting the $\log_{10}$ of each of the data points on regular graph paper, but semi-log paper simplifies this by permitting you to plot raw data and obtain the same line. Time is plotted on the linear horizontal x axis. **Draw a straight best-fit line through the data points.** The cells are growing logarithmically; therefore, the data should generate a straight line on semi-log paper (figure 10.3).

7. Also plot the data from the other temperature setting by averaging the class data on the blackboard.

8. Determine the generation time for *Escherichia coli* at each temperature. This can be done by arbitrarily selecting a point on the line and noting the O.D. Find the point on the line where this number has doubled. The time between these two points is the generation time.

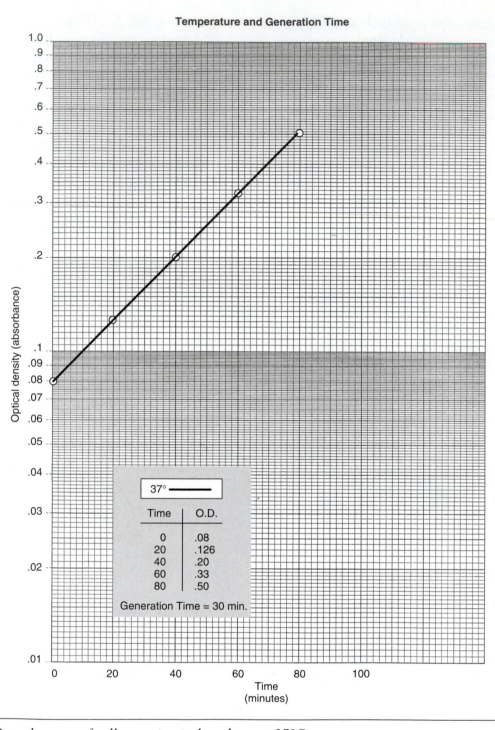

Temperature and Generation Time

37° ———	
Time	O.D.
0	.08
20	.126
40	.20
60	.33
80	.50

Generation Time = 30 min.

Figure 10.3 Growth curve of cells growing in log phase at 37°C.

EXERCISE

10

Laboratory Report: The Effect of Incubation Temperature on Generation Time

Results

Data: Your Temperature _____   Class Average Temperature _____

	Time	Reading	Time	Reading
1				
2				
3				
4				
5				

Generation Time

E. coli at 37°C

E. coli at 30°C

Questions

1. Which organism is growing faster: one with a short generation time or one with a longer generation time?

2. Why is it important to keep the culture at the correct incubation temperature when measuring the generation time?

3. If you didn't have semi-log paper, how could you plot the data on a piece of graph paper that would have resulted in a straight line?

4. If the growth of two cultures, one slower than the other, were plotted on semi-log paper, which would have the steeper slope?

5. What is the relationship between the density of the cells and the O.D.?

6. Where would you place the human population on the growth curve?

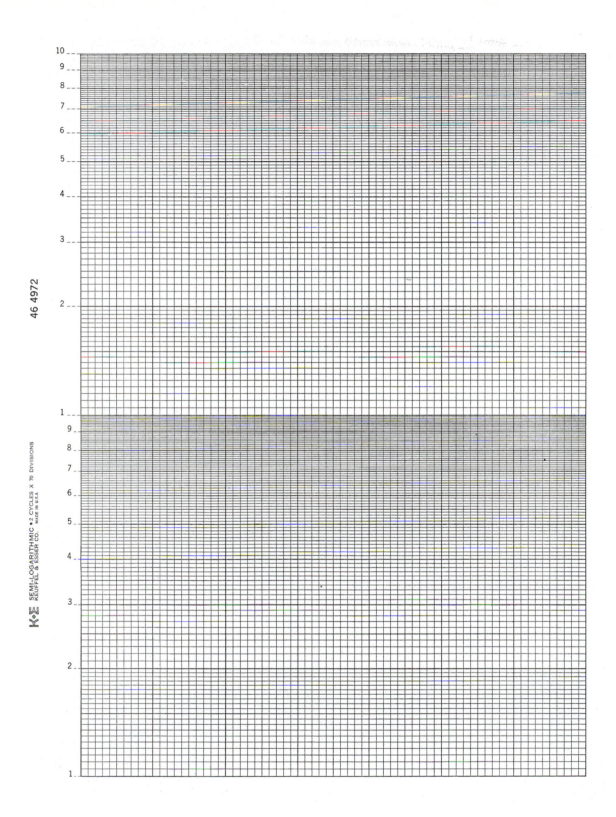

46 4972

K+E SEMI-LOGARITHMIC • 2 CYCLES X 70 DIVISIONS
KEUFFEL & ESSER CO. MADE IN U.S.A.

INTRODUCTION to Control of Microbial Growth

For many microbiologists, control of microbial growth means *maximization* of microbial growth—for example, when producing baker's yeast or in the production of antibiotics. To others, such as physicians and allied members of the medical profession, control means *minimization* of microbial growth—for example, the use of heat and ultraviolet light to destroy microorganisms present in growth media, gloves, and clothing. It can also imply the use of antiseptics, disinfectants, and antibiotics to inhibit or destroy microorganisms present on external or internal body parts.

Historically, Louis Pasteur (1822–1895) contributed to both areas. In his early research, he discovered that beer and wine making entailed a fermentation process involving initial growth of yeast. Later, he demonstrated that a sterile broth infusion in a swan-necked flask showed no turbidity due to microbial growth (figure I.5.1), and that upon tilting the flask, the sterile infusion became readily contaminated. The swan-necked flask experiment was both classical and monumental in that it helped resolve a debate of more than 150 years over the possible origin of microorganisms by spontaneous generation (abiogenesis).

The debate was finally squelched by John Tyndall, a physicist, who was able to establish an important fact overlooked by Pasteur—namely, that some bacteria in hay infusions existed in two forms: a **vegetative form,** readily susceptible to death by boiling of the hay infusion, and a resting form, now known as an endospore, which could survive boiling. With this knowledge, Tyndall developed a physical method of sterilization that we now describe as **tyndallization,** whereby both vegetative cells and endospores are destroyed when the infusion is boiled intermittently with periods of cooling. For sterilization of some materials by tyndallization, temperatures below boiling are possible. Tyndallization, although a somewhat lengthy sterilization method, is sometimes used to sterilize chemical nutrients subject to decomposition by the higher temperatures of autoclaving.

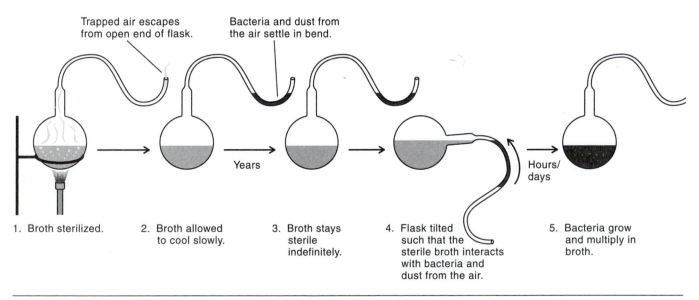

Trapped air escapes from open end of flask.

Bacteria and dust from the air settle in bend.

Years

Hours/days

1. Broth sterilized.

2. Broth allowed to cool slowly.

3. Broth stays sterile indefinitely.

4. Flask tilted such that the sterile broth interacts with bacteria and dust from the air.

5. Bacteria grow and multiply in broth.

Figure I.5.1 Pasteur's experiment with the swan-necked flask. If the flask remains upright, no microbial growth occurs (*1–3*). If microorganisms trapped in the neck reach the sterile liquid, they grow (*4 and 5*).

At about this same time, chemical disinfectants that aided in healing compound bone fractures were introduced by John Lister, an English surgeon, who was also impressed with Pasteur's findings. Lister had heard that carbolic acid (phenol) had remarkable effects when used to treat sewage in Carlisle; it not only prevented odors from farmlands irrigated with sewage, but it also destroyed intestinal parasites that usually infect cattle fed on such pastures.

The control of microbial growth has many applications today, both in microbiology and in such areas as plant and mammalian cell culture. Traditional examples of it include pure culture isolation and preparation of sterile culture media, bandages, and instruments. It also includes commercial preparation of various microbial products, such as antibiotics, fermented beverages, and food.

Exercises 7, 9, and 10 touched on maximization of microbial growth. In this section, the exercises deal with minimization or elimination of microbial growth by heat, ultraviolet light, osmotic pressure, antiseptics, and antibiotics.

… we are too much accustomed to attribute to a single cause that which is the product of several, and the majority of our controversies come from that.

Von Liebig

11

Moist and Dry Heat Sterilization: Thermal Death Point and Thermal Death Time

Getting Started

Physical methods used to kill microorganisms include heat and ultraviolet light. Filtration removes microorganisms from growth media, water, and from the air in operating rooms and cell transfer rooms; and using sterile gloves, masks, and clothing helps control air convection of microorganisms present on skin and hair. Ultraviolet light is used to kill microbes on surfaces like biological safety cabinet surfaces.

In this exercise, heat sterilization effects are studied because heat is commonly used to sterilize hospital, laboratory, industrial, and food items. When heat is applied, most microbes are killed, whereas cold temperatures inhibit but do not kill microbes. The sensitivity of a microorganism to heat is affected by its environment and genetics. Environmental factors include incubation temperature, chemical composition of the growth medium, and the age and concentration of cells in the growth medium. Genetically, **thermophiles** like *Bacillus stearothermophilus* and *Thermus aquaticus* grow at higher temperatures than others. In addition, *Bacillus* and *Clostridium* species can produce heat-resistant endospores.

Heat is applied in either a dry or moist form. Dry heat produced by hot-air ovens is used for sterilizing glassware like petri dishes and pipets, and micro-incinerators are used for sterilizing needles and loops. Dry heat kills by dehydrating microorganisms, resulting in **irreversible denaturation** of essential enzyme systems. Sterilization with dry heat requires more time and higher temperatures than with moist heat, because dry heat penetrates to the inside of microbial cells slower than moist heat. Typical sterilization times and temperatures are 2 hours at 165°C for dry heat compared to 15 minutes at 121°C at 15 psi (pounds per square inch of pressure) for moist heat. The mode of action is the same for both.

Autoclaving is the most commonly used moist heat sterilization method (color plate 10). Uses include sterilizing microbial media, cotton gowns, and some metal instruments. Other moist heat methods are **pasteurization, boiling,** and **tyndallization.** With autoclaving and tyndallization, both the vegetative and endospore forms of microorganisms are killed, whereas with pasteurization and boiling, usually only vegetative cells are killed. Boiling for 10 minutes is used to rid solutions such as drinking water of vegetative forms of pathogenic bacteria and other pathogens such as parasitic worms and protozoa.

Pasteurization, named for Louis Pasteur, is a moist heat process used to kill pathogenic bacteria and reduce the number of nonpathogenic bacteria such as **thermoduric** bacteria in beverages like milk, beer, and wine. The beverages are heated under controlled conditions of temperature and time, either 63°C for 30 minutes or 72°C for 15 seconds. (The lower temperatures of pasteurization help preserve food flavor.) However, many endospore-forming bacteria survive pasteurization.

Tyndallization, named after physicist John Tyndall, is sometimes used to sterilize nutrient media inactivated by the high temperature of autoclaving. The solution to be sterilized is usually steamed for 30 minutes in flowing steam (100°C) on each of three successive days. Between steaming times, the solution is left at room temperature. In principle, the first boiling kills all vegetative cells, the second boiling destroys newly germinated endospores, and the third boiling serves as added insurance that no living cells remain in the solution.

Two methods for determining the heat sensitivity of a microorganism are the **thermal death point (TDP)** and the **thermal death time (TDT).** The TDP is defined as the lowest temperature necessary to kill all of the microorganisms present in a culture in 10 minutes. The TDT is defined as the minimal time necessary to kill all of the microorganisms present in a culture held at a given temperature. These general principles are commonly used when establishing sterility requirements for various processes. Examples include milk, food preservation, and hospital supplies.

Definitions

Autoclave. A form of moist heat sterilization, conventionally performed at 121°C for 15 minutes at 15 psi.

Boiling. Moist heat treatment that kills vegetative forms of most pathogens. Conventionally, it is performed at 100°C; the time varies, although it is often done for 30 minutes.

Dry heat oven. Dry heat sterilization, conventionally done at 160°C–170°C for 2–3 hours.

Irreversible denaturation. A change in physical form of certain chemicals, which when heated, permanently destroys them. Enzymes constitute one such example.

Pasteurization. The use of moist heat at a temperature sufficiently high to kill pathogens but not necessarily all organisms. It is commonly used for milk products and is usually performed at 63°C for 30 minutes or 72°C for 15 seconds.

Thermal death point (TDP). The lowest temperature at which all microbes in a culture are killed after a given time.

Thermal death time (TDT). The minimal time necessary to kill all microbes in a culture held at a given temperature.

Thermoduric bacteria. Microbes able to survive conventional pasteurization, usually 63°C for 30 minutes or 72°C for 15 seconds.

Thermophile. An organism able to grow at temperatures above 55°C.

Tyndallization. The process of using repeated cycles of heating and incubation to kill spore-forming bacteria.

Vegetative form. Cells growing and dividing in typical fashion and not forming endospores.

Objectives

1. Describe physical sterilization methods requiring either moist heat (autoclaving, boiling, tyndallization, and pasteurization) or dry heat.
2. Explain the susceptibility of different bacteria to the lethal effect of moist heat—thermal death point (TDP) and thermal death time (TDT).

3. Appropriately use laboratory equipment commonly used for physical sterilization of moist and dry materials: the steam autoclave and the **dry heat oven.**

Prelab Questions

1. Name the genus and species of a bacterium that is a thermophile.
2. What is the difference between thermal death point and thermal death time?
3. Name the proper specifications of autoclaving.

Materials

Cultures of the following:

> 24-hour 37°C *Escherichia coli* cultures in 5 ml trypticase soy broth

> Spore suspension in 5 ml of sterile distilled water of a 4- to 5-day 37°C nutrient agar slant culture of *Bacillus subtilis*

5 ml of trypticase soy broth, 5 tubes per student

Two large beakers or cans, one for use as a water bath and the other for use as a reservoir of boiling water (per student)

Either a ring stand with wire screen and a Bunsen burner, a hot plate, or a heat block (per student)

One thermometer (per student)

A community water bath, or several water baths, one each per temperature

A vortex apparatus (if available)

PROCEDURE

First Session: Determination of Thermal Death Point and Thermal Death Time

Note: A procedural culture and heating distribution scheme for 8 students is shown in table 11.1. For this scheme, each student receives one broth culture, either *Escherichia coli* or *Bacillus subtilis*, to be heated at one of the assigned

Table 11.1 Culture and Heating Temperature Assignments (8 Students)

Bacterial Culture Assignment	Water Bath Temperature Assignment			
	40°C	55°C	80°C	100°C
Escherichia coli	1	2	3	4
Bacillus subtilis	5	6	7	8

temperatures (40°C, 55°C, 80°C, or 100°C). Students will label trypticase soy broth tubes for 0, 10, 20, 30, and 40 minutes.

Note: As an alternative, instead of each student preparing his or her own water bath, the instructor can provide four community water baths, preset at 40°C, 55°C, 80°C, and 100°C.

1. Suspend your assigned culture by gently rolling the tube between your hands, followed by aseptically transferring a loopful to a fresh tube of broth (label "0 time Control").

2. Fill the beaker to be used as a water bath approximately half full with water, sufficient to totally immerse the broth culture without dampening the test tube cap.

3. Place your tube of broth culture in the water bath along with an open tube of uninoculated broth in which a thermometer has been inserted for monitoring the water bath temperature.

4. Place the water bath and contents on either the ring stand or hot plate and heat almost to the assigned temperature. One or two degrees before the assigned temperature is reached, remove the water bath from the heat source and place it on your benchtop. The temperature of the water bath can now be maintained by periodically stirring in small amounts of boiling water obtained from the community water bath.

   Note: Students with the 100°C assignment may wish to keep their water bath on the heat source, providing the water can be controlled at a low boil.

5. After 10 minutes of heating, resuspend the broth culture either by vortexing or by gently tapping the outside of the tube. Aseptically transfer a loopful to a fresh tube of broth (label "10 minutes").

6. Repeat step 5 after 20, 30, and 40 minutes.

7. Write the initials of your culture as well your initials on all 5 tubes and incubate them in the 37°C incubator for 48 hours.

Demonstration of the Steam-Jacketed Autoclave and Dry Heat Oven

The Steam-Jacketed Autoclave

Note: As the instructor demonstrates the special features of the autoclave, follow the diagram in figure 11.1. Note the various control valves—their function and method of adjustment (exhaust valve, chamber valve, and safety valve); the steam pressure gauge; the thermometer and its location; and the door to the chamber. Also make note of the following special precautions necessary for proper sterilization.

1. Material sensitivity. Certain types of materials, such as talcum powder, oil, or petroleum jelly, cannot be steam sterilized because they are water repellent. Instead, dry heat is used. Some materials are destroyed (milk) or changed (medications) by the standard autoclave temperature of 121°C. In such instances, the autoclave may be operated at a lower pressure and temperature for a longer period of time. A heat-sensitive fluid material can normally be sterilized by filtration. Keep in mind that the smaller the filter pores, the slower the rate of filtration.

2. Proper preparation of materials. Steam must directly contact all materials to be sterilized. Therefore, media container closures such as metal caps with air passages, loosened screw cap lids, aluminum foil (heavy grade), and sometimes nonabsorbant cotton plugs are used. A small piece of autoclave tape or an indicator

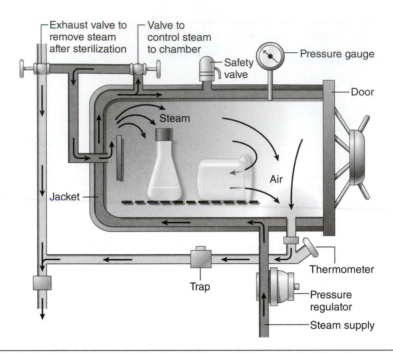

Figure 11.1 Autoclave. Steam first travels in an enclosed layer, or jacket, surrounding the chamber. It then enters the autoclave, displacing the air downward and out through a port in the bottom to the chamber.

vial should be included in each autoclave run to ensure sterility.

3. Proper loading of supplies. There must be ample space between packs and containers so that the steam can circulate. When using cotton plugs, they should be loosely covered with foil to prevent moisture from the autoclave condensing on them during cooling.

4. Complete evacuation of air from the chamber. This is necessary before replacement with steam. Older models may require manual clearance, but in newer models it is automatic.

5. Proper temperature. Autoclaving at a pressure of 15 psi achieves a temperature of 121°C (250°F) at sea level. If the temperature gauge does not register this reading, trapping of colder air in the autoclave is indicated, which lowers the temperature.

6. Adequate sterilization time. After the chamber temperature reaches 121°C, additional time is required for the heat to penetrate the material. The larger the size of individual containers and packs, the more time required. Time must be adjusted to the individual load size.

7. Completion of the autoclaving process. Rapid reduction in steam pressure can cause fluids to boil vigorously through the caps or to explode. Drop steam pressure gradually as cooling occurs and, when possible, allow material to dry in the autoclave. If removed while moist, the wrappings or plugs may provide a means for reentry of bacteria present in the room air.

8. To ensure items were completely autoclaved, check to see that the autoclave tape turned black. Alternatively, an autoclaved indicator vial may need to be incubated at 37°C for 24–48 hours.

The Dry Heat Oven

The hot air, or dry heat, oven is used in most laboratories for both drying glassware and sterilization.

When using the dry heat oven, the following guidelines are important:

1. Materials suitable for sterilization include oil; petroleum jelly; Vaseline; metal containers; and dry, clean glassware.

2. An oven with circulating air takes about half the sterilization time of a static air oven.

Better heat transfer occurs with circulating air (i.e., convection ovens).

3. Proper packaging is necessary to ensure air circulation to the inside surfaces. For example, syringes must be separate from the plunger so that all surfaces are exposed to circulating air.

4. Sterilization of dirty materials should be avoided. The presence of extraneous materials such as protein delays the process and may allow bacteria to survive inside the material.

5. In part 2 of the Laboratory Report, prepare a list of materials for your class that are sterilized in the autoclave. For each one, indicate the standard temperature, pressure, and time used for sterilization.

Do the same for materials sterilized in the dry heat oven. Indicate the oven sterilization temperature, time, and reasons for sterilizing there.

Second Session

After 48 hours, examine your broth tubes for the presence or absence of turbidity (growth). Write your results in the appropriate place in the table on the blackboard of your classroom. When all the results are entered, transfer them to table 11.2 of the Laboratory Report.

EXERCISE

11

Laboratory Report: Moist and Dry Heat
Sterilization: Thermal Death Point
and Thermal Death Time

Results

1. Determination of thermal death point and thermal death time:

Table 11.2   Bacterial Growth at Assigned Temperatures and Times

Culture	40°C					55°C					80°C					100°C				
	C	10	20	30	40	C	10	20	30	40	C	10	20	30	40	C	10	20	30	40
Escherichia coli																				
Bacillus subtilis																				

Note: C = Control; 10, 20, 30, 40 = minutes of heating the inoculated culture at the assigned temperature; Use a + sign for growth and a − sign for no growth

a. Determine the thermal death time for each culture.

Thermal Death Time (Minutes)

Escherichia coli _____

Bacillus subtilis _____

b. Determine the thermal death point for each culture.

Thermal Death Point (°C)

Escherichia coli _____

Bacillus subtilis _____

2. Plot a graph of the thermal death time.

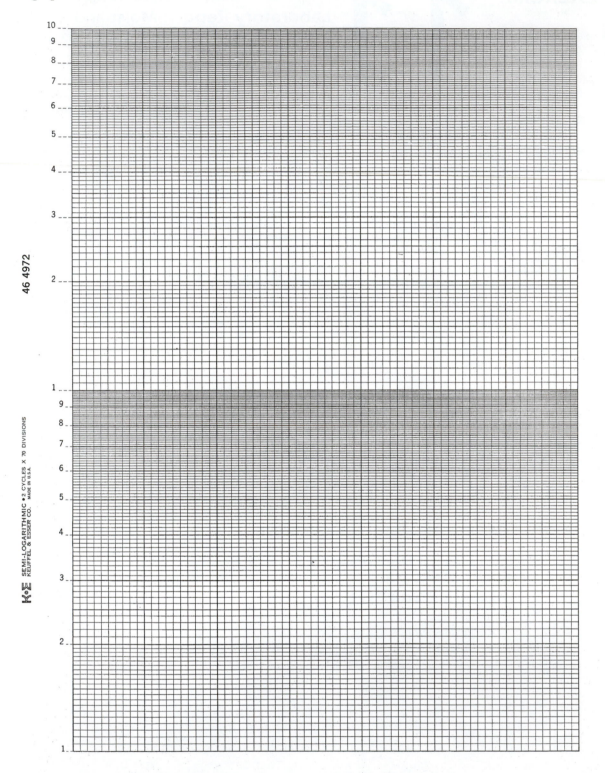

46 4972

K•E SEMI-LOGARITHMIC • 2 CYCLES X 70 DIVISIONS
KEUFFEL & ESSER CO.   MADE IN U.S.A.

3. Evaluation of materials sterilized in our laboratory with moist heat (autoclave) and dry heat (hot air oven):

   a. List of materials sterilized with the autoclave (see Procedure for criteria):

   b. List of materials sterilized with the dry heat oven (see Procedure for criteria):

Questions

1. Discuss similarities and differences between determining thermal death point and thermal death time.

2. How would you set up an experiment to determine to the minute the TDT of *E. coli*? Begin with the data you have already collected.

3. A practical question related to thermal death time (TDT) relates to a serious outbreak of *E. coli* infection in early 1993 when people ate insufficiently grilled hamburgers. How would you set up an experiment to determine the TDT of a solid such as a hamburger? Assume that the thermal death point (TDP) is 67.2°C (157°F), the temperature required by many states to cook hamburger on an open grill. What factors would you consider in setting up such an experiment?

4. What is the most expedient method for sterilizing a heat-sensitive liquid that contains a spore-forming bacterium?

5. List one or more materials that are best sterilized by the following processes:
   a. Membrane filtration

   b. Ultraviolet light

   c. Dry heat

   d. Moist heat

   e. Tyndallization

6. What are three advantages of using metal caps rather than cotton for test tube closures? Are there any disadvantages?

7. How would you sterilize a heat-sensitive growth medium containing thermoduric bacteria?

8. Was the *Bacillus subtilis* culture sterilized after 40 minutes of boiling? If not, what is necessary to **ensure** sterility by boiling?

12

Control of Microbial Growth with Ultraviolet Light

Getting Started

Ultraviolet (UV) light is the component of sunlight that is responsible for sunburn. It can also kill microorganisms by acting on their DNA, causing mutations. It consists of very short wavelengths of radiation (175–350 nm) located just below blue light (450–500 nm) in the **visible spectrum** (figure 12.1).

The actual mechanism of mutation is the formation of **thymine dimers** (figure 12.2). Two adjacent **thymines** on a DNA strand bind to each other; and when the DNA is replicated, an incorrect base pair is frequently incorporated into the newly synthesized strand. This may cause mutations, and if there is sufficient radiation, ultimately, the death of the cell.

UV light does not penetrate surfaces and will not go through ordinary plastic or glass. It is only useful for killing organisms *on* surfaces and in the air. Sometimes UV lights are turned on in operating rooms and other places where airborne bacterial contamination is a problem. Because UV light quickly damages the eyes, these lights are turned on only when no one is in the irradiated area.

Bacteria vary in their sensitivity to UV light. In this exercise, the sensitivity of *Bacillus* **endospores** will be compared with non-spore-forming cells. You will also irradiate a mixed culture, such as organisms in soil or hamburger, to compare the resistance of

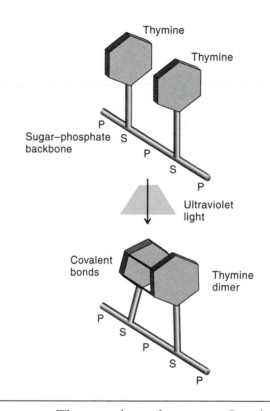

Figure 12.2 Thymine dimer formation. Covalent bonds form between adjacent thymine molecules on the same strand of DNA. This distorts the shape of the DNA and prevents replication of the changed DNA.

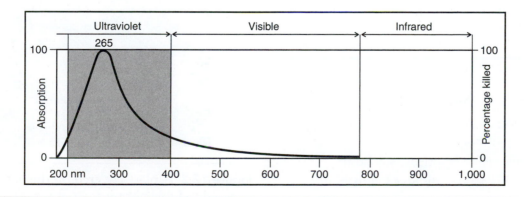

Figure 12.1 Germicidal activity of radiant energy.

different organisms. Immediately following irradiation, it is important to place the samples in a dark container because some cells potentially killed by UV light recover when exposed to longer wavelengths of light (**light repair**).

Definitions

Endospores. A resting stage formed by some organisms that is resistant to heat, drying, chemicals, and starvation.

Light repair. DNA repair in cells previously exposed to UV light by a DNA repair enzyme that requires visible light.

Thymine. A pyrimidine: one of the four nucleotide subunits of DNA.

Thymine dimer. A molecule formed when two adjacent thymine molecules in the same strand of DNA covalently bond to one another.

Ultraviolet (UV) light. Electromagnetic radiation with wavelengths between 175 and 350 nm; invisible.

Visible spectrum. Wavelengths of radiation between 400 and 800 nm.

Objectives

1. Explain the effects of ultraviolet (UV) light irradiation on bacteria.
2. Explain how to work safely with ultraviolet light.

Prelab Questions

1. Why are safety glasses required in this exercise when turning on the UV light?
2. What component of the cell is affected by UV light?
3. Why was *Bacillus* chosen as one of the cultures to irradiate?

Materials

Cultures of the following:

Suspension of *Bacillus* endospores in sterile saline

Escherichia coli in trypticase soy broth

Raw hamburger or soil mixed with sterile water

Trypticase soy agar plates, 3 per team

UV lamp with shielding. An 18- to 36-inch fluorescent bulb with a wavelength of 240–280 nm is ideal. It enables uniform exposure of 3–6 partially opened petri dishes.

Sterile swabs, 3

Safety glasses for use with the UV lamp

Dark box for storing plates after UV exposure

PROCEDURE

First Session

Safety Precautions: (1) The area for UV irradiation should be in an isolated part of the laboratory. (2) Wear safety glasses as a precautionary measure when working in this area. (3) Never look at the UV light after turning it on because it could result in severe eye damage. Skin damage is also a slight possibility.

1. Dip a sterile swab in a suspension of *Bacillus* spores and swab an agar plate in three directions as shown in figure 14.2.
2. Repeat the procedure with a suspended *E. coli* broth culture.
3. For the third plate, you can either dip the swab into a mixture of sterile water and hamburger or sterile water and soil.

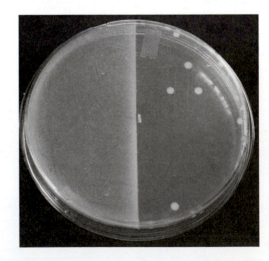

Figure 12.3 A plate covered on the left and exposed to UV light on the right. © Anna Oller, University of Central Missouri

4. Place the plates under a UV lamp propped up about 20 cm from the bench surface. Open the petri dish(es) and partially cover each plate with the lid. The part of the plate protected by the lid will be the control because UV radiation does not penetrate most plastic.

5. Put on your safety glasses and turn on the UV light. Expose the plates to UV radiation for 3 minutes. (Your instructor may assign different exposure times to some teams.)

6. Turn off the UV light. Cover the plates, invert them, and place them in a covered container. Incubate at 37°C for 48 hours.

Second Session

Observe the plates and record your findings in the Laboratory Report.

EXERCISE

12

Laboratory Report: Control of Microbial Growth with Ultraviolet Light

Results

1. Record your observations for control and treated sides of petri dishes exposed to UV light. Record your results on the blackboard.

	Assigned Exposure	Optional Class Exposures
Bacillus endospores		
Escherichia coli		
Soil or meat suspension		

Score irradiated growth:

++++ no different than nonirradiated control

+++ slightly different than nonirradiated control

++ significantly less growth than nonirradiated control

+ less than 10 colonies

2. Make a drawing of each plate
    a. Plate containing *Bacillus* spores

    b. Plate containing *E. coli*

    c. Plate containing either a raw hamburger suspension or soil suspension (indicate which one you used)

3. Which organisms were most resistant to UV light? _____ least resistant?_____

Questions

1. Why can't you use UV light to sterilize microbiological media, *e.g.*, agar or broth?

2. How does UV light cause mutations?

3. What is the result of increasing the time of exposure to UV light?

4. Give a possible reason some organisms in the soil (or meat) were able to grow after exposure to UV light but others were not.

5. Frequently organisms isolated from the environment are pigmented, while organisms isolated from the intestine or other protected places are not. Can you provide an explanation for this?

6. Mutations can lead to cancer in animals. Explain why persons living in the southern half of the United States have a higher incidence of skin cancer than those in the northern half.

EXERCISE

13

Changes in Osmotic Pressure Due to Salt and Sugar and Their Effects on Microbial Growth

Getting Started

Osmosis, which is derived from the Greek word "alter," refers to the process of flow or diffusion that takes place through a **semipermeable membrane.** In a living cell, the cytoplasmic membrane, located adjacent to the inside of the cell wall, represents such a membrane (figure 13.1*a*).

Both the cytoplasmic membrane and the cell wall help prevent the cell from either bursting (**plasmoptysis**) (figure 13.1*a*) or collapsing (**plasmolysis**) (figure 13.1*b*) due to either entry or removal of water from the cell, respectively. The **solute** concentration both inside and outside the cell determines which, if any, process happens. When the solute concentration inside the cell is the same as the solute concentration on the outside of the cell (**isotonic**), the cell remains intact. When the solute concentration outside the cell is less than the solute concentration inside the cell, an inward **osmotic pressure** occurs, and water enters the cell in an attempt to equalize the solute concentration on either side of the cytoplasmic membrane. If the solute concentration outside the cell is sufficiently low (**hypotonic**), the cell will absorb water and sometimes burst (plasmoptysis). However, this rarely occurs, due to the rigidity of the cell wall. The reverse phenomenon, cell shrinkage followed by cell lysis (plasmolysis), can occur when the cell is placed in a more concentrated (**hypertonic**) solution. This can become a life-and-death problem if too much water is removed from the cells (figure 13.1*b*).

When placed in an isotonic solution, some cells recover, although there are many genera that die once the external osmotic pressure exceeds their limitations. This concept is the basis of food preservation— the use of high salt concentrations (for cheese and pickle brine) and high sugar concentrations (in honey and jams).

In general, fungi (yeasts and molds) are much more resistant to high external solute concentrations

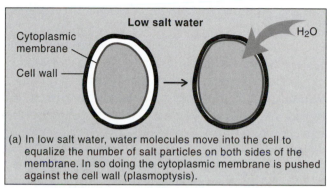

(a) In low salt water, water molecules move into the cell to equalize the number of salt particles on both sides of the membrane. In so doing the cytoplasmic membrane is pushed against the cell wall (plasmoptysis).

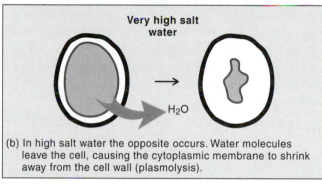

(b) In high salt water the opposite occurs. Water molecules leave the cell, causing the cytoplasmic membrane to shrink away from the cell wall (plasmolysis).

Figure 13.1 Movement of water into and out of cells. (*a*) Low and (*b*) very high salt-containing solutions. The cytoplasmic membrane is semipermeable and only allows water molecules to pass through freely.

than are bacteria, which is one reason fungi can grow in or on jelly, cheese, and fruit. There are, of course, exceptions among bacteria; for example, the genus *Halobacterium* in *Archaea* is found in nature growing in water with a high salt content, like Great Salt Lake in Utah; and the genus *Micrococcus halobius* in the family *Micrococcaceae* is sometimes found in nature growing on highly salted (25–32%) protein products, such as fish and animal hides. It is interesting to note that all of these bacteria produce a red pigment.

There are also some salt-tolerant *Staphylococcus* strains able to grow at salt concentrations greater than 10% (w/v), enabling them to grow on skin surfaces. The salt-loving (halophilic) bacteria are also unique in that, like all *Archaea*, they lack muramic acid as a bonding agent in their cell walls. Instead, their cell walls are believed to contain sodium and potassium ions. These ions help confer cell wall rigidity, perhaps helping explain the reason they require such high salt concentrations for good growth.

Yeasts and molds able to grow in high sugar (50–75%) and sometimes high salt (25–30%) concentrations are termed **saccharophilic** and halophilic fungi, respectively. Some of these yeast and fungus genera are *Saccharomyces*, *Aspergillus*, and *Penicillium*.

In this exercise, you will examine the ability of some of the previously mentioned halophiles and saccharophiles to grow on the surface of TYEG (trypticase yeast extract glucose) agar plates containing increasing concentrations of salt and sugar. *Escherichia coli* is added as a salt-sensitive Gram-negative rod control. You will also have an opportunity to examine any changes in cell form with increasing salt and sugar concentrations.

Keep in mind that all of these halophilic and saccharophilic microbes are characterized by an increase in lag time and a decrease in growth rate and in the amount of cell substance synthesized. In some ways, their **growth curve** (Exercise 10) parallels what happens when they are grown at a temperature below their optimal growth temperature. For example, halobacteria have a **generation time** of 7 hours and halococci have one of 15 hours.

Definitions

Generation time. The time required for one cell to divide into two cells.

Growth curve. A curve describing the four readily distinguishable phases of microbial growth: lag, log, stationary, and death.

Halophilic microbe. A salt-requiring organism able to grow in a medium containing a salt concentration high enough to inhibit other organisms.

Hypertonic fluid. A fluid having an osmotic pressure greater than another fluid with which it is compared.

Hypotonic fluid. A fluid having an osmotic pressure lower than another fluid with which it is compared.

Isotonic fluid. A fluid having the same osmotic pressure as another fluid with which it is compared.

Osmotic pressure. The pressure exerted by water on a membrane as a result of a difference in the concentration of solute molecules on each side of the membrane.

Plasmolysis. Contraction or shrinking of the cytoplasmic membrane away from the cell wall due to a loss of water from the cell.

Plasmoptysis. The bursting of protoplasm from a cell due to rupture of the cell wall when absorbing excess water from the external environment.

Saccharophilic microbe. An organism able to grow in environments containing high sugar concentrations.

Semipermeable membrane. A membrane, such as the cytoplasmic membrane of the cell, that permits passage of some materials but not others. Passage usually depends on the size of the molecule.

Solute. A dissolved substance in a solution.

Objectives

1. Describe osmotic pressure and show how it can be used to inhibit growth of less osmotolerant microbes while allowing more osmotolerant microbes to grow, although often at a considerably slower growth rate.

2. Interpret if microorganisms either require or grow better in an environment containing high concentrations of salt (halophilic) or sugar (saccharophilic).

Prelab Questions

1. If a cell is placed in a hypotonic solution, what will happen to the cell?

2. What is the proper term given to a microbe that is able to grow in a high salt concentration?

3. What growth differences between the lowest and highest salt concentration do you hypothesize you will see for *Escherichia coli*?

Materials

Per team

Cultures of the following:

Trypticase yeast extract glucose (TYEG) agar slants for *Escherichia coli, Micrococcus luteus (or Staphylococcus aureus)*, and *Saccharomyces cerevisiae*

American Type Culture Collection (ATCC) medium 213 for the Preceptrol strain of *Halobacterium salinarium*

Incubate *E. coli, M. luteus* (or *S. aureus*), and *S. cerevisiae* cultures for 24 hours at 35°C; incubate *H. salinarium* culture for 1 week (perhaps longer) at 35°C

TYEG agar plates containing 0.5, 5, 10, and 20% salt (NaCl), 4

TYEG agar plates containing 0, 10, 25, and 50% sucrose, 4

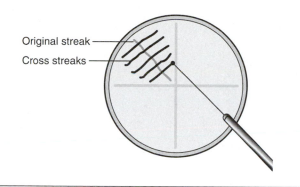

Figure 13.2 Streaking procedure for isolation of single colonies.

PROCEDURE

First Session

Note: One student of the pair can inoculate the four trypticase soy agar plates containing 0.5, 5, 10, and 20% NaCl, while the other student inoculates the four plates containing 0, 10, 25, and 50% sucrose.

1. First, use a glass-marking pen to divide the bottom of the eight plates in quadrants, and label with the initials of the four test organisms; for example, *E.c.* for *Escherichia coli*, and so on. Also label the underside of each plate with the salt or sugar concentrations and your name.

2. Using aseptic technique, remove a loopful from a culture and streak the appropriate quadrant of your plate in a straight line approximately 1 inch long. Then reflame your loop, cool it for a few seconds, and make a series of cross streaks approximately 1/2 inch long in order to initiate single colonies for use in studying colonial morphology (figure 13.2). Repeat the inoculation procedure used for the first culture in the appropriate quadrant of the remaining three agar plates.

3. Repeat the inoculation procedure for the remaining three test organisms.

4. Invert and incubate the eight plates at 30°C.

5. Observe the plates periodically (up to 1 week or more if necessary) for growth.

Second Session

1. Examine your plates for the presence (+) or absence (−) of growth. For growth, use one to three + signs (+ = minimal, + + = some, and + + + = good). Enter results in table 13.1 (various salt concentrations) and table 13.2 (various sugar concentrations) of the Laboratory Report.

2. Compare the colonial growth characteristics of cultures grown on agar media containing increasing salt and sugar concentrations. Make notes of any marked changes in colony color, colony size (in mm), and colony texture: dull or glistening, rough or smooth, and flat or raised. Record your findings in the Laboratory Report.

3. Prepare wet mounts of bacteria and yeast colonies showing marked changes in visual appearance with increasing salt and sugar concentrations. Examine bacteria and yeasts with the high–dry objective lens. Look for plasmolyzed cells and other changes, such as cell form and size. Prepare drawings of any such changes in the Laboratory Report.

EXERCISE

13

Laboratory Report: Changes in Osmotic Pressure Due to Salt and Sugar and Their Effects on Microbial Growth

Results

1. Examination of petri dish cultures for the presence (+) or absence (−) of growth in the presence of increasing salt (table 13.1) and sugar concentrations (table 13.2). Use a series of one to three + signs to describe the amount of growth.

Table 13.1   Presence or Absence of Growth on TYEG Agar Plates Containing NaCl and Incubated for 48 Hours—1 Week

| Culture | NaCl Concentration (%) | | | | | | | |
| | 0.5 | | 5 | | 10 | | 20 | |
	48 hr	1 wk	48 hr	1 wk	48 hr	1 wk	48 hr	1 wk
Escherichia coli								
Halobacterium salinarium								
Micrococcus luteus/Staphylococcus aureus								
Saccharomyces cerevisiae								

Table 13.2   Presence or Absence of Growth on TYEG Agar Plates Containing Sucrose and Incubated for 48 Hours—1 Week

| Culture | Sucrose Concentration (%) | | | | | | | |
| | 0 | | 10 | | 25 | | 50 | |
	48 hr	1 wk	48 hr	1 wk	48 hr	1 wk	48 hr	1 wk
Escherichia coli								
Halobacterium salinarium								
Micrococcus luteus/Staphylococcus aureus								
Saccharomyces cerevisiae								

2. Comparison of the colonial growth characteristics of cultures inoculated on agar media containing increasing amounts of salt or sugar.
   a. *Escherichia coli*

NaCl %	Growth	Colony Color	Colony Size	Colony Texture
0.5				
5				
10				
20				

Sucrose %	Growth	Colony Color	Colony Size	Colony Texture
0				
10				
25				
50				

b. *Halobacterium salinarium*

NaCl %	Growth	Colony Color	Colony Size	Colony Texture
0.5				
5				
10				
20				

Sucrose %				
0				
10				
25				
50				

c. *Micrococcus luteus (or Staphylococcus aureus)*

NaCl %	Growth	Colony Color	Colony Size	Colony Texture
0.5				
5				
10				
20				

Sucrose %	Growth	Colony Color	Colony Size	Colony Texture
0				
10				
25				
50				

d. *Saccharomyces cerevisiae*

NaCl %	Growth	Colony Color	Colony Size	Colony Texture
0.5				
5				
10				
20				

Sucrose %

0				
10				
25				
50				

3. Examine wet mounts of bacteria and yeast colonies showing *marked* changes in appearance from the control colonies. How do the cells appear?

Questions

1. From your studies, which organism(s) tolerate salt best? _____ least? _____
2. Which organism(s) tolerate sugar best? _____ least? _____
3. Compare bacteria and yeast with respect to salt tolerance. Bear in mind both colonial and cellular appearance in formulating your answers.

4. Compare bacteria and yeast with respect to sugar tolerance. Bear in mind both colonial and cellular appearance in formulating your answer.

5. What evidence did you find of a *nutritional requirement* for salt or sugar in the growth medium?

6. Matching
    Each answer may be used one or more times.
    a. *Halobacterium*   _____ osmosensitive
    b. *Saccharomyces*   _____ long generation time
    c. *Escherichia coli*   _____ saccharophilic
    d. *Micrococcus/Staphylococcus* _____ osmotolerant

7. Matching
    Choose the best answer. Each answer may be used one or more times or not at all.
    a. Plasmolysis   _____ isotonic solution
    b. Plasmoptysis   _____ hypotonic solution
    c. Normal cell growth _____ hypertonic solution
                           _____ swelling of cells

EXERCISE

14

Antibiotics

Getting Started

Chemicals such as antibiotics have been in existence a long time. However, the therapeutic properties of antibiotics simply were not recognized until Alexander Fleming's discovery of penicillin in the 1930s.

By definition, **antibiotics** are chemicals produced and secreted by microorganisms, especially bacteria and fungi, that can inhibit the growth of or kill, other microorganisms. While some of the antibiotics in use today are naturally produced, such as penicillin, the majority of antibiotics prescribed are either semisynthetic or completely synthetic. Semisynthetic antibiotics are made by chemically modifying a naturally made antibiotic, such as amoxicillin derived from penicillin. Ciprofloxacin and the sulfa drugs represent synthetic antibiotics; these are completely synthesized in a chemistry lab.

The first chemicals used as antibiotics were the synthetic sulfa drugs, such as sulfanilamide, which originated from the azo group of dyes (figure 14.1). The inhibitory action of the sulfa drugs is termed **competitive inhibition,** in which the sulfanilamide acts as an **antimetabolite.** Antimetabolites are molecules that resemble essential cellular compounds but block cellular function when incorporated into the cell's metabolism. The sulfanilamide replaces para-aminobenzoic acid (PABA), an essential metabolite required in the biosynthetic pathway to generate folic acid. In the presence of sulfanilamide, bacterial cells cannot synthesize folic acid and cannot grow. This drug is selectively toxic to bacterial cells and not human cells because it inhibits a pathway necessary to generate folic acid. Humans obtain folic acid from dietary sources such as dark green leafy vegetables, beans, and meat.

The Kirby-Bauer test is still used in many clinical laboratories to test the potency of antibiotics and drugs. In this assay, discs of filter paper are impregnated with antibiotic solutions containing the same range of concentrations that would be found dissolved in the tissues of the human body. The agar plate is

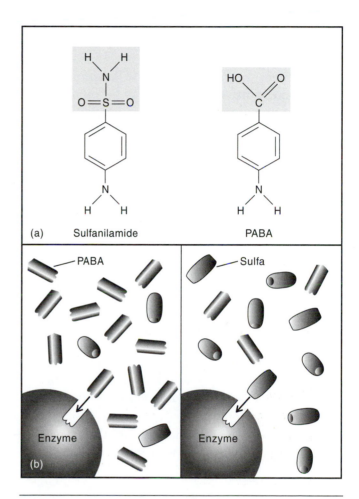

Figure 14.1 (*a*) Structures of sulfanilamide (sulfa drug) and of para-aminobenzoic acid (PABA). The portions of the molecules that differ from each other are shaded. (*b*) Reversible competitive inhibition of folic acid synthesis by sulfa drug. The higher the concentration of sulfa drug molecules relative to PABA, the more likely that the enzyme will bind to the sulfa drug, and the greater the inhibition of folic acid synthesis.

inoculated as a lawn of growth with bacteria, and the discs are then placed onto the agar plate surface. When incubated, the bacteria grow in a smooth lawn of confluent growth, except in the area around the

Exercise 14 Antibiotics 14–1

103

antibiotic disc, if the organism is **susceptible** to the drug. This clear zone formed around the antibiotic disc is called the **zone of inhibition.** It is important to note that the presence of a zone of inhibition does not necessarily indicate that the organism is sensitive to the antibiotic. Some antibiotics are smaller molecules and diffuse faster producing a larger zone. Susceptibility to a particular antibiotic is determined by measuring the diameter of the zone of inhibition and referencing the antibiotic susceptibility chart. In a clinical lab, the Kirby-Bauer test is performed using special conditions, such as 2- to 5-hour cultures, controlled inoculum size, and short incubation periods. Because these conditions are difficult to achieve within the time frame of most classrooms, a modified Kirby-Bauer test will be performed.

In this lab exercise, you will perform a modified Kirby-Bauer test to determine the antibiotic sensitivity of four bacterial organisms (see Materials section for names of organisms). You will need to reference the antibiotic standard chart (table 14.1) to determine if the organism is **resistant,** intermediate, or sensitive **(susceptible)** to the antibiotic in question.

Definitions

Antibiotic. A chemical produced largely by certain bacteria and fungi that can inhibit or destroy the growth of other organisms, including pathogenic microorganisms.

Antimetabolite. A substance that inhibits the utilization of a metabolite necessary for growth.

Coenzyme. Any heat-stable, nonprotein compound that forms an active portion of an enzyme system after combination with an enzyme precursor (apoenzyme). Many of the B vitamins are coenzymes.

Competitive inhibition. The inhibition of enzyme activity by competition between the inhibitor and the substrate for the active (catalytic) site on the enzyme.

Essential metabolic pathway. A pathway of chemical transformations necessary for growth; if inhibited, the organism usually dies. The Krebs cycle and Embden-Meyerhof pathway are classic examples.

Essential metabolite. A chemical necessary for proper growth.

Pathogen. Any agent capable of causing disease, usually a microorganism.

Resistant. In relation to antibiotics, a bacterium not killed by the chemical compound.

Susceptible. In relation to antibiotics, a bacterium killed by the chemical compound.

Zone of inhibition. Region around a chemical saturated disc where bacteria are unable to grow due to adverse effects of the compound in the disc.

Objectives

1. Describe the origin and use of antibiotics.
2. Describe the mechanism of action of the first synthetic antibiotics, the sulfa drugs.
3. Assess the inhibitory activity of antibiotics on bacteria using a modified Kirby-Bauer test.

Table 14.1 Chart Containing Zone Diameter Interpretive Standards for Determining the Sensitivity of Bacteria to Antimicrobial Agents

Antimicrobial Agent	Disc Content	Susceptible	Intermediate	Resistant
Penicillin G when testing	10U			
Staphylococci and				
Streptococci		≥29	None	≤28
when testing Enterococci		≥15	None	≤14
Streptomycin	10 µg	≥15	12–14	≤11
Tetracycline	30 µg	≥19	15–18	≤14
Ciprofloxacin	5 µg	≥21	16–20	≤15
Trimethoprim & sulfamethoxazole*	1.25 + 23.75 µg	≥16	11–15	≤10
Erythromycin	15 µg	≥23	14–22	≤13

Source: BBL. Used by permission.
* Note: Sulfamethoxazole (a sulfa drug) is usually given in combination with trimethoprim.

Prelab Questions

1. What is the definition of *antibiotic*?

2. What was the first synthetic antibiotic? What was the mechanism of action?

3. How can you determine if an antibiotic is an effective therapeutic agent for a patient?

Materials

Cultures (per team of 2–4 students) of the following:

Bacteria (24-hour 37°C TS broth cultures)

   Staphylococcus epidermidis (a Gram-positive coccus)

   Escherichia coli (a Gram-negative rod)

   Pseudomonas aeruginosa (a nonfermenting Gram-negative rod)

   Mycobacterium smegmatis (an acid-fast rod)

Vials or dispensers of the following antibiotic discs:

   Penicillin, 10 µg; streptomycin, 10 µg; tetracycline, 30 µg; ciprofloxacin, 5 µg; sulfanilamide (or another sulfonamide), 300 µg; erythromycin, 15 µg

Mueller–Hinton agar, 4 large plates OR use 8 regular size plates

Sterile cotton swabs, 4

Small forceps, 1 per student

Ruler divided in mm

PROCEDURE

First Session: Filter Paper Disc Technique for Antibiotics

1. Divide the four broth cultures among team members so that each student sets up at least one susceptibility test.

2. With a permanent pen, divide the underside of four large plates of Mueller–Hinton agar into six pie-shaped sections (figure 14.2a). If using standard size petri plates, divide the plates into three sections, placing three of the six antibiotics on each plate.

3. Record the codes of the six antibiotic discs on the bottom side of the four plates, one code for each section (table 14.2 of the Laboratory Report has code designations).

4. Label the bottom of each plate with the name of the respective bacterium (Materials section has names).

5. Using aseptic technique, streak the first broth culture with a sterile swab in horizontal and vertical directions and around the edge of the agar plate (as shown in figure 14.2b) to create a lawn of growth. The remaining three cultures should be streaked on separate plates in a similar manner.

6. Heat sterilize forceps by dipping them in 95% alcohol and then touch them to the flame of the Bunsen burner. Air cool the forceps and remove

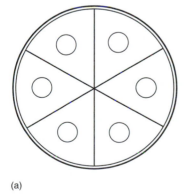

(a)

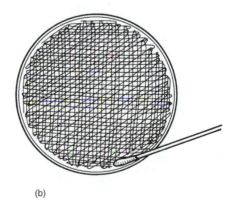

(b)

Figure 14.2 Antibiotic susceptibility test. (*a*) The underside of a Mueller–Hinton agar plate showing the marking of sections and the arrangement for placement of antibiotic discs on the agar surface. (*b*) Procedure for streaking an agar plate in three or more directions with a swab inoculum in order to achieve a uniform lawn of growth (see Figure 12.3).

an antibiotic disc from the container. Place gently, with identification side up, in the center of one of the pie-shaped sections of the agar plate (figure 14.2a). Tap gently to fix in position.

7. Continue placing the remaining five antibiotic discs in the same way.

   Note: Be sure to flame the forceps with alcohol after placing each disc because it is possible to contaminate stock vials with resistant organisms.

   Note: If a disc dispenser is used, follow the manufacturer's instructions.

8. Repeat steps 5–7 with the remaining three cultures.

9. Invert and incubate the plates at 37°C for 48 hours.

Second Session: Filter Paper Disc Technique for Antibiotics

1. On the bottom of the *Staphylococcus epidermidis* plate, and with a ruler calibrated in mm, determine the diameter of the clear zone (zone of inhibition) surrounding each disc. Repeat this process with the three other bacterial plates. In addition, make note of any small colonies present in the clear zone of inhibition surrounding each antibiotic disc as they may be resistant mutants.

   Note: It may be necessary to illuminate the plate in order to define the boundary of the clear zone.

2. Record your findings in table 14.2 of the Laboratory Report.

3. Compare your results where possible with table 14.1 and indicate in table 14.2 the susceptibility of your test cultures (*when possible*) to the antibiotics as resistant (R), intermediate (I), or susceptible (S).

   Note: Your answers may not agree exactly with those in table 14.1 because this is a modified Kirby-Bauer test.

EXERCISE

14

Laboratory Report: Antibiotics

Results

Table 14.2   Antibiotic Susceptibility (Modified Kirby-Bauer Test)

Antibiotic	Cip		Ery		Pen		Str		Sul		Tet	
	Zone (mm)	S/I/R	Zone (mm)	S/I/R	Zone (mm)	S/I/R	Zone (mm)	S/I/R	Zone (mm)	S/I/R	Zone (mm)	S/I/R
S. epidermidis												
E. coli												
M. smegmatis												
P. aeruginosa												

R = Resistant, I = Intermediate, and S = Susceptible.

Questions

1. What relationship did you find, if any, between the Gram-staining reaction of a microorganism and its susceptibility to an antibiotic?

2. Did you observe any small colonies growing in the zone of inhibition around the antibiotic disc? Explain how this might occur.

3. When performing the Kirby-Bauer procedure to evaluate a clinical specimen, why is it important to have a standardized inoculum size, 2–5 hour cultures, and a short incubation period?

4. Compare your antibiotic sensitivity results for the antibiotic penicillin with *E. coli* and *S. epidermidis*. Why is there a difference in sensitivity?

5. You test a clinical specimen of *Streptococcus* with the antibiotics penicillin and tetracycline. The zones of inhibition for penicillin and tetracycline are both 20 mm. What does this tell you about comparing zone sizes and the use of the antibiotic zone diameter chart? Is this organism sensitive to both of these antibiotics?

15

Antiseptics and Disinfectants

Getting Started

Joseph Lister, the pioneer of aseptic surgery, recognized early in the 1860s that the use of chemicals such as carbolic acid (phenol) dramatically reduced the incidence of post-surgical infections. Since this time, the use of antiseptics and disinfectants in hospitals has been essential to control infection and limit the spread of disease.

Antiseptics are chemicals able to inhibit **sepsis** (infection). They do not kill the sepsis-producing agent; they merely inhibit its growth. Antiseptic chemicals must be sufficiently nontoxic to allow application to skin and mucous membranes, such as the use of Listerine® for gargling. These same

chemicals may also act as **disinfectants** (chemicals able to kill vegetative forms but not necessarily kill endospore forms of bacteria) when used at higher concentration levels. Toxicity is a major factor in determining usage of a chemical as either an antiseptic or disinfectant. While antiseptics are used on the skin and mucous membranes, disinfectants are used on inanimate surfaces, also known as **fomites.**

There are many different classes of chemicals used in the medical field as antiseptics or disinfectants (table 15.1). Alcohols are typically used as antiseptics to cleanse the skin before injections or as an integral component of tinctures. Bisphenols, derived from phenolics, are chemicals found to function as

Table 15.1   Chemical Compounds Commonly Used in Hospitals for Controlling Growth of Microorganisms

Sodium hypochlorite (5%)	Disinfectant	External surfaces, such as tables
Iodine (1% in 70% alcohol)	Disinfectant	External surfaces, such as tables
Iodophors (70 ppm avail. I_2)	Disinfectant	External surfaces, such as tables
Quaternary ammonium compounds	Disinfectant	External surfaces, such as tables
Phenol (5%), carbolic acid, source coal tar	Disinfectant	External surfaces, such as tables
Hexachlorophene (pHisoHex®)	Disinfectant	Presurgical hand washing
Formaldehyde (4%)	Disinfectant	Oral and rectal thermometers
Iodophors (70 ppm avail. I_2)	Disinfectant	Oral and rectal thermometers
Alcohol, ethanol (70%)	Antiseptic	Skin
Iodine (tincture in alcohol with KI)	Antiseptic	Skin
Iodophors	Antiseptic	Skin
Organic mercury compounds (merthiolate, mercurochrome)	Antiseptic	Skin
Hydrogen peroxide (3%)	Antiseptic	Superficial skin infections
Potassium permanganate	Antiseptic	Urethral and superficial skin fungus infections
Silver nitrate (1%)	Antiseptic	Prevention of eye infections in newborn babies
Zinc oxide paste	Antiseptic	Diaper rash
Zinc salts of fatty acids (Desenex®)	Antiseptic	Athlete's foot
Glycerol (50%)	Antiseptic	Prevent bacterial growth in stool and surgical specimens
Ethylene oxide gas (12%)	Sterilization	Linens, syringes, etc.
Formaldehyde (20% in 70% alcohol)	Sterilization	Metal instruments
Glutaraldehyde (pH 7.5 or more)	Sterilization	Metal instruments

either disinfectants or antiseptics. Orthophenylphenol, a bisphenol, is the active ingredient found in the disinfectant Lysol. Another bisphenol known as triclosan is found in products such as toothpaste, deodorant, and mouthwashes. It is also currently incorporated into consumer products such as garbage bags, diapers, and cutting boards. Heavy metal compounds such as zinc and silver can be found in household products as well as medical products. Zinc in the form of zinc oxide is found as the key ingredient in sunscreens and diaper rash creams. Silver in the form of silver nanoparticles is used as an antiseptic and can be found in topical antiseptic gels, bandages (Curad® brand), and surgical mesh dressings. Samsung has created a line of products called Silver Nano®, which includes silver nanoparticles integrated in the surface of household appliances.

Choosing the right chemical for a particular situation is more complex than it seems. There are many factors that influence the effectiveness of antiseptics and disinfectants such as: the pH and temperature of the environment, concentration of the antiseptic or disinfectant, presence of organic matter, and type of organism targeted. For example, many chemicals work optimally at a neutral pH and the effectiveness of the chemical is often reduced if the product is used in an acidic or alkaline environment. An increase in temperature during disinfection also increases the rate at which microbes are destroyed. Organic material will often inhibit the activity of many chemicals such as hypochlorite, whereas phenolics remain effective in their presence. **Endospores, non-enveloped viruses,** and *Mycobacteria* are the most resistant to general-purpose disinfectants and often require high-level disinfectants or sterilants such as ethylene oxide and glutaraldehyde to destroy them. It is also important to note that high-level sterilants, such as glutaraldehyde, are also very toxic.

Most of these chemicals inhibit microbial growth by altering the cell wall structures, damaging the cytoplasmic membrane (quaternary ammonium compounds), or denaturing essential proteins (alcohols, heavy metals) such as enzymes. If the proteins or lipids that are part of the cell wall structure are compromised, the cell will no longer retain its shape and cell lysis occurs when placed in a hypotonic solution. Damage to the cytoplasmic membrane will impact the ability of the cell to transport nutrients in and waste out. Membranes that are part of the envelope of viruses are also very sensitive to chemicals, and if altered, hinder the ability of the virus to bind to target cells. This is why enveloped viruses are more sensitive to chemicals than non-enveloped viruses. When proteins are denatured, the hydrogen or disulfide bonds between amino acids that maintain their three-dimensional shape are broken. This results in a protein that ceases to function. Some chemicals exhibit one of the above mechanisms as well as lower the surface tension, as seen in antibacterial soaps.

In this lab, you will test the effectiveness of various antiseptics and disinfectants with the bacterial organisms listed in the materials section. Filter paper discs are soaked with representative test chemicals, then placed on the surface of TSA plates covered with an organism. The plates are incubated at 37°C for 24–48 hours and then evaluated for the presence of clear zones of inhibition around the discs containing the test chemicals. To compare the effectiveness of the test chemicals toward different organisms, you will measure the diameter of the zones of inhibition.

Definitions

Antiseptic. A chemical that inhibits or kills microbes. The definition also implies that the chemical is sufficiently nontoxic that it can be applied to skin and mucous membranes.

Disinfectant. A chemical agent that rids an area of pathogenic microorganisms. In so doing, it kills vegetative forms of bacteria but ordinarily not endospore forms. The definition also implies that the chemical is sufficiently toxic that it should not be applied to body surfaces, only to material objects.

Endospore. A resting stage formed by some organisms that is resistant to heat, drying, chemicals, and starvation.

Enveloped virus. A virus with an envelope surrounding the protein coat. The HIV virus is an enveloped virus.

Fomite. Inanimate objects such as books, tools, or towels that can act as transmitters of pathogenic microorganisms or viruses.

Non-enveloped virus. A virus without an envelope surrounding the protein coat, often called a naked virus. The hepatitis type A virus is a non-enveloped virus.

Sepsis. The presence of pathogenic microorganisms or their toxins in tissue or blood; infection.

Objectives

1. Describe the differences between an antiseptic and disinfectant.
2. Describe the factors that influence the effectiveness of chemicals and choose the appropriate chemical for a particular situation.
3. Assess the effectiveness of a chemical to inhibit microbial growth.

Prelab Questions

1. When would you use an antiseptic? When would you use a disinfectant?
2. List one example of how chemicals are able to inhibit bacterial growth.
3. In this lab, how will you determine if an organism is sensitive to a chemical?

Materials

Cultures (per team of 2–4 students) of the following:

Bacteria (24-hour 37°C trypticase soy broth cultures)

Staphylococcus epidermidis (a Gram-positive coccus)

Escherichia coli (a Gram-negative rod)

Pseudomonas aeruginosa (nonfermenting Gram-negative rod)

Bacillus cereus endospores in saline (**Note:** best from 5-day-old culture)

Beakers containing 10 ml aliquots of the following chemicals: 70% ethanol, 3% hydrogen peroxide, antiseptic mouthwash such as Listerine®, and disinfectant such as Lysol®

Trypticase soy agar plates, 4

Sterile cotton swabs, 4

Small forceps, 1 per student

Ruler divided in mm

PROCEDURE

First Session: Filter Paper Disc Technique for Antiseptics and Disinfectants

1. With a glass-marking pen, divide the underside of four plates of trypticase soy agar into quadrants and label them 1–4.
2. Record codes for the four antiseptics and disinfectants on the bottom sides of the four agar plates, one code for each quadrant: 70% ethanol—E; 3% hydrogen peroxide—HP; Listerine®—L; and Lysol®—Ly.
3. Label the bottom of each plate with the name of the respective bacterium (Materials section has names).
4. Suspend the *Staphylococcus epidermidis* culture, insert and moisten a sterile swab in the culture, remove excess, then streak the swab in three directions on the surface of the agar plate to create a lawn of growth. Discard swab in the appropriate waste container.
5. Repeat step 4 with the other cultures used in this lab.
6. Sterilize forceps by dipping them in 95% alcohol and then touch them to the flame of the Bunsen burner. Air cool.
7. Using forceps, remove one of the filter paper discs from the container and dip it into solution 1—70% ethanol.
8. Drain the disc thoroughly on a piece of clean absorbent toweling and place it in the center of quadrant 1 of the dish labeled *Staphylococcus epidermidis* (figure 15.1). Tap disc gently.
9. Repeat steps 5, 6, and 7 and place the disc in the center of quadrant 1 of the three other plates.

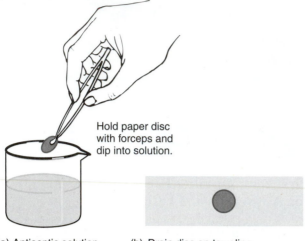

Hold paper disc with forceps and dip into solution.

(a) Antiseptic solution

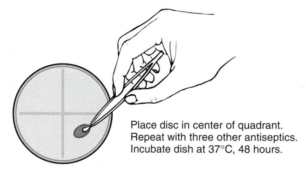

(b) Drain disc on toweling.

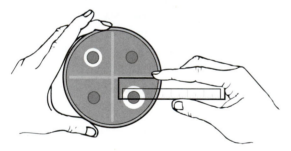

Place disc in center of quadrant. Repeat with three other antiseptics. Incubate dish at 37°C, 48 hours.

(c) Petri dish seeded with *S. aureus* or *E. coli*

(d) Measure the clear zone of inhibition surrounding each disc.

Figure 15.1 Filter paper disc technique for antiseptics and disinfectants.

10. Repeat steps 5–8 for the remaining three compounds, using first 3% hydrogen peroxide, then the antiseptic mouthwash, and last Lysol®.

11. Invert the petri dishes and incubate at 37°C for 48 hours.

Second Session: Filter Paper Disc Technique for Antiseptics and Disinfectants

1. Turn over the *Staphylococcus epidermidis* plate, and with a ruler calibrated in mm, determine the diameter of the clear zone (zone of inhibition) surrounding each disc. Repeat this process with the three other plates.

2. The presence of a zone of inhibition reflects the organism's susceptibility to that chemical. Record your results in table 15.2 of the Laboratory Report.

Note: It may be necessary to illuminate the plate in order to define the clear zone boundary.

EXERCISE

15

Laboratory Report: Antiseptics and Disinfectants

Results

Zones of inhibition are measured in millimeters.

Table 15.2   **Inhibitory Activity of Various Antiseptics and Disinfectants**

Antiseptic or Disinfectant	Zone of Inhibition (mm)			
	Staphylococcus epidermidis	Bacillus cereus endospores	Escherichia coli	Pseudomonas aeruginosa
70% ethanol (E)				
3% hydrogen peroxide (HP)				
Listerine® (L)				
Lysol® (Ly)				
Others:				

What general conclusions can you make from this study? What differences, if any, did you observe on your plates between antiseptics and disinfectant preparations?

Questions

1. What relationship did you find, if any, between the Gram-staining reaction of a microorganism and its susceptibility to antiseptics and disinfectants?

2. Were any of the organisms especially resistant to the chemicals tested? If so, which one(s)? Can you propose a reason for these results?

3. Based on your test results, which of the test chemicals is the most effective antiseptic? Which of the test chemicals is the most effective disinfectant?

4. When choosing a chemical for a particular application, what should be considered?

5. What type of chemical (list an example of a specific chemical) would you use to
   a. disinfect the skin before an injection? _____
   b. disinfect lab benchtops before/after class? _____
   c. disinfect the skin before surgery? _____
   d. disinfect drinking water? _____
   e. destroy *Bacillus anthracis* endospores contaminating a room? _____

INTRODUCTION to Microbial Genetics

In this section, three aspects of microbial genetics will be studied: selection of mutants, gene transfer, and gene regulation.

Selection of Mutants Mutations are constantly occurring in all living things. The replication of DNA is amazingly error-free, but about once in every 100 million duplications of a gene, a change is made. There are three possible outcomes:

1. There is no effect. Perhaps the altered base did not lead to structural change in a protein and the cell remained functional.

2. The mutation has affected a critical portion of an essential protein, resulting in the death of the cell.

3. The mutation has enabled the cell to grow faster or survive longer than the other nonmutated cells. This outcome rarely occurs.

Gene Transfer Bacteria can transfer genetic material to other bacteria in three ways. These are conjugation, transduction, and transformation.

Conjugation occurs during cell-to-cell contact and is somewhat similar to sexual recombination seen in other organisms. The transferred DNA can be either chromosomal or a small, circular piece of DNA called a plasmid.

Transduction is the transfer of genes from one bacterial cell to another by a bacterial virus. These viruses, called bacteriophage or phage, package a bacterial gene along with the viral genes and transfer it to a new cell.

The third method of transferring genes is transformation, which is also called DNA-mediated transformation. (The word *transformation* is sometimes used to define the change of normal animal cells to malignant cells—a completely different system.) In bacterial transformation, isolated DNA is mixed with viable cells. It then enters the cells, which are able to express these new genes. Although it seems impossible for a large molecule such as DNA to enter through the cell wall and membrane of a living cell, this is indeed what happens.

Gene Regulation Another aspect of genetics is the expression of genes. A cell must be economical with its energy and material and must not make enzymes or other products when they are not needed. On the other hand, a cell has to be able to "turn on" genes when they are required in a particular environment. Gene regulation is examined in exercise 18.

Notes:

EXERCISE 16

Selection of Bacterial Mutants Resistant to Antibiotics

Getting Started

All the bacterial cells in a pure culture are de-rived from a single cell. These cells, however, are not identical because all genes tend to mutate and form mutant organisms. The spontaneous **mutation rate** of genes varies between 1 in 10^4 to 1 in 10^{12} cell divisions, and even though they are quite a rare event, significant mutations are observed because bacterial populations are very large. In a bacterial suspension of 10^9 cells/ml, one could expect 10 mutations of a gene that mutated 1 in every 10^8 divisions.

Mutant bacteria usually do not grow as well as the **wild-type** normal cell because most changes are harmful, or at least not helpful. If, however, conditions change in the environment and favor a mutant cell, it will be able to outcompete and outgrow the cells that do not have the advantageous mutation. It is important to understand that the mutation is a random event that the cell cannot direct. No matter how useful a **mutation** may be in a certain situation, it just happens to the cell, randomly conferring an advantage or disadvantage to it.

In this exercise, you will select bacteria resistant to streptomycin. Streptomycin is an **antibiotic** that kills bacteria by acting on their ribosomes to prevent protein synthesis. (However, it does not stop protein synthesis in animals because eukaryotic ribosomes are larger than those of bacteria and therefore different.) **Sensitive** E. coli cells can become resistant to streptomycin with just one mutation.

In this exercise, you will select organisms resistant to streptomycin by adding a large population of sensitive bacteria to a bottle of trypticase soy broth containing streptomycin. Only organisms that already had a random mutation for streptomycin resistance will be able to survive and multiply (figure 16.1).

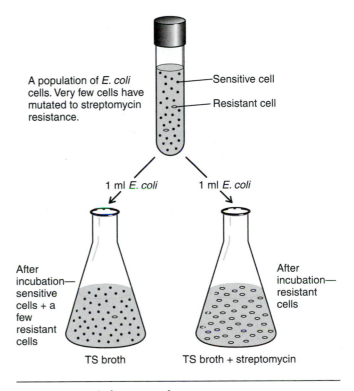

Figure 16.1 Selection of streptomycin-resistant E. coli cells.

Definitions

Antibiotic. A substance produced by one organism, usually a microorganism, that kills or inhibits other organisms.

Mutation. An inheritable change in the base sequences of DNA.

Mutation rate. The number of mutations per cell division.

Sensitive. An organism killed or inhibited by a particular antibiotic.

Wild type. The organism as it is isolated from nature.

Objectives

1. Explain the concept of selection and its relationship to mutation.

2. Explain that mutations are random events and that the cell cannot cause specific mutations to occur, no matter how advantageous they may be.

3. Calculate the number of streptomycin-resistant mutant bacteria that occur in an overnight culture of a sensitive strain.

Materials

Per team

First Session

    Flasks (or bottles) containing 50 ml trypticase soy broth, 2

    Trypticase soy agar deeps, 2

    Sterile petri dishes, 2

    1 ml pipets, 2

    Overnight broth culture (~18 hours) of *Escherichia coli* K12 (about 10^9 cells/ml)

    Streptomycin solution at 30 mg/ml

Second Session

    Trypticase soy agar deeps, 2

    Sterile petri dishes, 2

    1 ml pipets, 2

    Tubes of 0.5 ml sterile water, 2

    0.1 ml streptomycin

PROCEDURE

First Session

1. Melt and place 2 trypticase soy agar deeps in a 50°C water bath.

2. Label one petri plate and one flask "with streptomycin." Label the other flask and plate "without streptomycin control" (figure 16.2).

3. Add 0.3 ml streptomycin to the flask labeled "with streptomycin" and 0.1 ml to one of the melted, cooled agar deeps. Discard the pipet.

4. Immediately inoculate the trypticase soy agar deep with 1 ml of the bacterial culture, mix, and pour in the plate labeled "with streptomycin." Only organisms that already have mutated to streptomycin resistance will be able to grow in this plate.

5. Add 1 ml of the bacteria to the tube of melted, cooled agar without streptomycin and pour into plate labeled "without streptomycin control."

6. Add 1 ml of bacteria to each of the flasks.

7. Incubate the plates and flasks at 37°C. If using bottles, lay them on their side to increase aeration.

Second Session

1. Melt and cool two tubes of trypticase soy agar in a 50°C water bath.

2. Pour one tube of melted agar into a petri dish labeled "without streptomycin" and let harden.

3. Add 0.1 ml streptomycin to the other tube of melted agar, pour into a petri dish labeled "with streptomycin," and let harden.

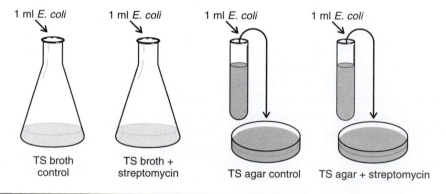

Figure 16.2 Inoculating media with and without streptomycin with a culture of *E. coli* (session 1).

4. Examine the bottles and plates inoculated during the first session. Note whether there is growth (turbidity) or not in both of the bottles. Count the number of colonies growing in the pour plates. How many streptomycin-resistant mutants/ml were present in the original inoculum? Compare it to the growth of organisms in the control plate without streptomycin. If there are more than 300 colonies or the plate is covered by confluent growth, record as TNTC or "too numerous to count." Record results.

5. Test the bacteria growing in the bottles and on the plates for sensitivity or resistance to streptomycin in the following way: Divide both agar plates in four sections as diagrammed in figure 16.3, then take a loopful of broth from the bottle without streptomycin and inoculate a sector of each agar plate. Do the same with the broth culture containing streptomycin.

6. Dig an isolated colony out of the agar plate containing the streptomycin and suspend it in a tube of sterile water. Use a loopful to inoculate the third sector of each plate. Also suspend some organisms from the control plate in sterile water (there will not be any isolated colonies) and inoculate the fourth sector. Incubate the plates at 37°C.

7. Predict which bacteria will be sensitive to streptomycin and which will be resistant.

Third Session

1. Observe the growth on each sector of the plates and record results. Were they as you predicted?

2. Occasionally, mutants not only will be resistant to streptomycin but also will require it. If you have one of these unusual mutants, be sure to show it to the instructor.

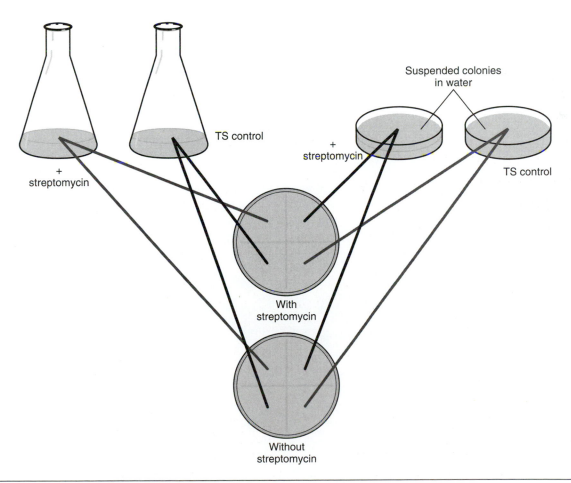

Figure 16.3 Testing previously incubated cultures for streptomycin sensitivity (session 2).

EXERCISE 16

Laboratory Report: Selection of Bacterial Mutants Resistant to Antibiotics

Results (Second Session)

Source	Growth / No Growth
TS broth (control)	
TS broth plus streptomycin	

Source	Number of Colonies
TS agar plate (control)	
TS agar plate plus streptomycin	

Results (Third Session)

Source	Growth on TS Agar Plate	Growth on TS Agar Plate + Strp
TS broth (control)		
TS broth plus streptomycin		
TS agar (control)		
TS agar plus streptomycin		

How many organisms/ml were streptomycin resistant in the original overnight culture of sensitive *E. coli*?

Questions

1. Two bottles of trypticase soy broth (with and without streptomycin) were inoculated in the first session with 1 ml of an overnight culture of *Escherichia coli*. After incubation, why was one population sensitive to streptomycin and the other population resistant?

2. How were you able to estimate the number of streptomycin-resistant organisms already present in the overnight culture of *Escherichia coli* growing in the trypticase soy broth?

3. Why should antibiotics only be used if they are necessary?

4. Which is correct?
   a. An organism becomes resistant after it is exposed to an antibiotic.

   b. An antibiotic selects organisms that are already resistant.

5. Why is streptomycin ineffective for treating viral diseases?

EXERCISE 17

Transformation: A Form of Genetic Recombination

Getting Started

In this exercise, transformation is used to transfer the genes of one bacterium to another. It gives you a chance to see the results of what seems to be an impossible process—a huge DNA molecule entering an intact cell and permanently changing its genetic makeup.

Basically, the process involves mixing DNA from one strain of lysed (disrupted) cells with another strain of living cells. The DNA then enters the viable cells and is incorporated into the bacterial chromosome. The new DNA is expressed, and the genetic capability of the cell may be changed.

In order to determine whether the bacteria are indeed taking up additional DNA, the two sets of organisms (DNA donors and DNA recipients) must differ in some way. One strain usually has a "marker," such as resistance to an antibiotic or the inability to synthesize an amino acid or vitamin. In this exercise, a gene responsible for conferring resistance to the antibiotic streptomycin is transferred to cells that are sensitive to it (figure 17.1).

The organism used in this exercise is *Acinetobacter* (a-sin-NEET-o-bacter), a short, Gram-negative rod found in soil and water. The prefix *a* means "without," and *cine* means "movement," as in cinema; thus *Acinetobacter* is nonmotile. This organism is always **competent,** which means it can always take up **naked DNA.** Most bacteria are not competent unless they are in a particular part of the growth curve or in a special physiological condition. In any event, for transformation to occur, the DNA must not be degraded (chemically broken down). If an enzyme such as **DNase** is present, it cuts the DNA in small pieces, preventing transformation.

Definitions

Competent. Cells that are able to take up naked DNA.

DNase. An enzyme that cuts DNA, making it useless for transformation.

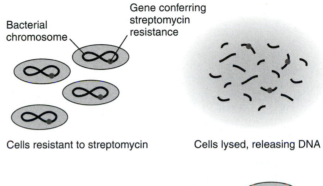

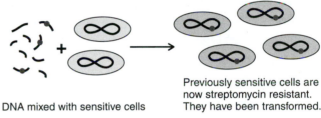

Figure 17.1 Transformation of cells with a gene conferring streptomycin resistance.

Lysate. The solution that contains cell contents like membranes, DNA, and proteins.

Naked DNA. DNA released from lysed, or disrupted, cells and no longer protected by an intact cell.

Objectives

1. Describe the process of transformation, and observe it in the laboratory.
2. Interpret the use of genetic markers.
3. Explain the importance of controls in an experiment.

Prelab Questions

1. Why is *Acinetobacter* used for the transformation instead of other genera of bacteria?
2. How do you lyse cells to release DNA?
3. What is the purpose of adding streptomycin to the TSY plate?

Safety Precaution: *Acinetobacter* can cause pneumonia in immunologically compromised individuals.

Materials

Per team

TSY (trypticase soy yeast extract) agar plate, 1

TSY agar plate with streptomycin (second session), 1

Broth culture of *Acinetobacter* Str^R (resistant to streptomycin), 1

Broth culture of *Acinetobacter* Str^S (sensitive to streptomycin), 1

Tube with 0.1 ml detergent SDS (sodium dodecyl sulfate) in 10× saline citrate, 1

Solution of DNase

1 ml pipet, 1

Class equipment

60°C water bath with test tube rack

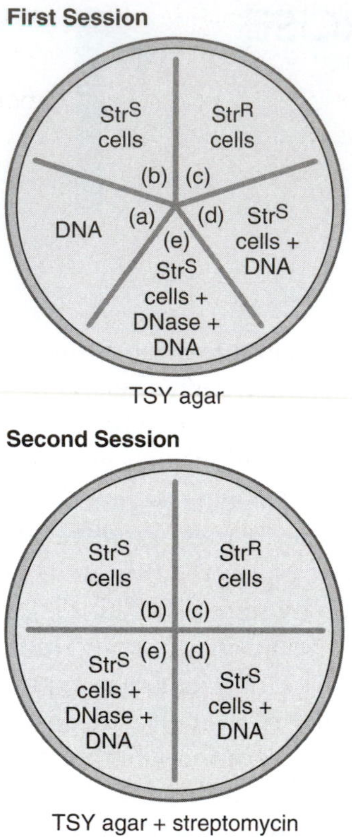

Figure 17.2 Five labeled sectors of a TSY agar plate.

PROCEDURE

First Session

1. Transfer 1.0 ml of Str^R *Acinetobacter* broth culture into the tube of SDS. Label the tube and incubate it in the 60°C water bath for 30 minutes. The detergent (SDS) will lyse the cells, releasing DNA and other cell contents into the solution. This is called the **lysate.** Any cells that are not lysed will be killed by the 30-minute exposure to 60°C water. Label the tube "DNA."

2. Divide the bottom of the trypticase soy yeast extract agar plate into five sectors using a marking pen. Label the sections "DNA," "Str^S (streptomycin sensitive)," "Str^R (streptomycin resistant)," "Str^S + DNA," and "Str^S + DNase + DNA" (figure 17.2).

3. Inoculate the TSY plate as indicated by adding a loop of the broth culture, DNA, or DNase in

an area about the size of a dime to each sector. Avoid cross-contamination of sectors.

a. DNA. The lysed mixture of Str^R cells is the source of DNA. It also contains RNA, proteins, and all the other cell components of the lysed cells that do not interfere with the transformation.

b. Str^S cells. Inoculate a loopful of the Str^S culture.

c. Str^R. Inoculate a loopful of the Str^R culture.

d. Str^S cells + DNA. Inoculate a loopful of Str^S cells and then add a loopful of the DNA (lysed Str^R cells) in the same area. THIS IS THE ACTUAL TRANSFOR-MATION. Str^R cells will grow here if transformed by the DNA.

e. Inoculate a loopful of Str^S cells as above, and in the same area add a loopful of DNase, then add a loopful of DNA. It is important to add these in the correct order or transformation will occur before the

DNase can be added. The DNase should degrade the DNA and prevent transformation.

4. Incubate the plates at room temperature for several days or at 37°C for 48 hours.

Second Session

1. Observe the plate you prepared in the first session. There should be growth in all sectors of the plate except the DNA sector (*a*). If the DNA control sector shows growth, it indicates that your crude DNA preparation was not sterile and contained viable cells. If this has happened, discard your plates and borrow another student's plate after he or she is finished with it; there should be sufficient material for more than one team. Why is it so important that the DNA preparation is sterile?

2. Be sure you understand the purpose of each control. Which sector demonstrates that

    a. the Str^S cells were viable and can grow on the agar plate?

    b. the Str^R cells were viable?

    c. the lysed mixture of Str^R cells contained no viable organisms?

    d. DNA is the component of the lysed cells responsible for transformation?

3. Divide the bottom of a TSY + streptomycin plate into four sectors and label them "Str^S," "Str^R," "Str^S + DNA," and "Str^S + DNase + DNA."

4. Streak a loopful of cells from the first plate to the corresponding sectors on the TSY + streptomycin plate. Lightly spread them in an area about the size of a dime. Cells growing on this plate must be streptomycin resistant.

5. Incubate at room temperature for several days or at 37°C for 48 hours or until cells have grown.

Third Session

1. Observe the TSY + streptomycin agar plate inoculated last session and record results. Did you transform the cells sensitive to streptomycin to cells that were resistant and could now grow on streptomycin?

2. Did the results from the controls confirm that naked DNA can transfer resistance to streptomycin?

EXERCISE 17

Laboratory Report: Transformation:
A Form of Genetic Recombination

Results

Indicate growth (+) or no growth (−) in each sector.

Results of first session

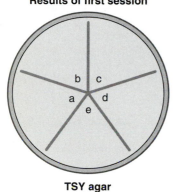

TSY agar

Results of second session

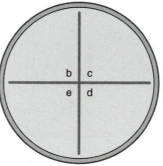

TSY agar + streptomycin

1. Were *Acinetobacter* Str^S cells sensitive to streptomycin?

2. Were *Acinetobacter* Str^R cells resistant to streptomycin?

3. Was the DNA (cell lysate) free of viable cells?

4. Did transformation take place?

5. Did the DNase prevent transformation?

Questions

1. What two components were mixed together to show transformation?

Yes/No	Plate and Sector That Demonstrate Observation

2. What is the action of DNase?

3. What control showed that transformation—not conjugation or transduction—was responsible for the results?

4. If the Str^S cells placed in section B had grown on the TSY + streptomycin agar plate, would you have been able to determine if transformation had taken place? Explain.

5. If you had used a DNA lysate containing viable cells, would it have been possible to determine whether transformation had taken place? Explain.

6. How does transformation differ from conjugation and transduction?

INTRODUCTION to the Other Microbial World

The phrase *the other microbial world* refers to organisms other than bacteria, the major organisms of study in other parts of the manual. All of the organisms included here, with the exception of viruses, are eukaryotic organisms, many of which are of medical importance. Included are members of the nonfilamentous fungi (primarily yeasts), the filamentous fungi (molds), and intestinal animal parasites of medical importance, the protozoa and helminths. Viruses that infect both prokaryotes (bacteria) and eukaryotes (animal and plant cells) are introduced. Only bacterial viruses will be made available for laboratory study.

Mycology, the study of fungi, is the subject of exercise 19. Fungi differ from algae in that they lack chlorophyll, and they differ from bacteria in that the cells are larger and contain membrane-bound organelles. Most yeasts and filamentous fungi found in nature are nonpathogenic. In fact, many contribute to our well-being: for example, the yeast *Saccharomyces* is important in manufacturing bread, beer, and wine; and the filamentous fungus *Penicillium chrysogenum* produces the antibiotic penicillin. Some cause spoilage (moldy jam and bread), and *Aspergillus* produces food toxins (aflatoxins). Many filamentous fungi, such as mushrooms, puffballs, toadstools, bracket fungi, and molds, are visible with the naked eye.

Antibiotics effective against prokaryotes are usually ineffective against eukaryotes. If an antibiotic is effective against a fungus, it may also, depending on its mode of action, damage the human host, because humans are also eukaryotic. Consequently, antibiotic control of fungal infections is usually more difficult than control of bacterial infections. Fortunately, many fungal infections are opportunistic[1] infections so that healthy individuals rarely acquire them, other than cutaneous fungal infections such as athlete's foot.

Fungi can be cultivated in the laboratory in the same manner as bacteria. Physiologically, all fungi are heterotrophs (they require an organic source of carbon, such as glucose). Most are aerobic, although some are facultative anaerobes, and a few are obligate anaerobes (found in the rumens of cows). Most fungi grow best at 20°–30°C, although some grow well at 45°–50°C (such as *Aspergillus fumigatus*, an opportunistic filamentous fungus known to cause pulmonary aspergillosis).

Parasitic diseases constitute a major worldwide public health problem, both in developed and developing countries. In developing countries, parasitic diseases are prevalent due to poverty, malnutrition, lack of sanitation, and lack of education. A simple family survey for intestinal parasites in a small Central American village revealed that every family member harbored at least three types of parasites. Effects of human parasitic diseases range from minimal, nutritional loss with minor discomfort (pinworm and *Ascaris* infections), to debilitating and life-threatening, such as malaria and schistosomiasis.

Industrialized countries also suffer from parasitic infections. Fecal contamination of drinking water by wild animal carriers, such as beavers, has caused major outbreaks of **giardiasis** (an intestinal disease) in various parts of the United States. The causative agent, the protozoan *Giardia lamblia* (see figure 29.3), produces **cysts** that are small, thereby enabling them to sometimes pass through faulty water supply filters. The protozoan flagellate *Trichomonas vaginalis* is a common cause of vaginitis in women and is often sexually transmitted. Pinworm infestations are a problem among elementary schoolchildren. About 20% of infections in domestic animals are caused by protozoan and helminthic (worm) agents.

Because of their worldwide public health importance and their natural history differences from bacteria and fungi, we believe protozoa and helminths merit their own laboratory session. Exercise 20 provides

---

[1] Opportunistic infections are associated with debilitating diseases (such as cancer) and use of cytotoxic drugs, broad-spectrum antibiotics, and radiation therapy, all of which can suppress the normal immune response.

insights on the diseases they cause and the techniques used for their diagnosis and identification.

Virology, the study of *viruses* (the word for "poison" in Greek), has early roots, although somewhat mysterious. Viruses could not be seen, even with a light microscope, and yet when sap from the leaf of an infected tobacco plant was passed through a filter that removed bacteria and fungi, the clear filtrate retained its infectious properties. It was not until the mid-1930s, with the advent and aid of the electron microscope, that viruses were first observed. In 1935, Wendell Stanley succeeded in crystallizing tobacco mosaic virus (TMV), enabling him to observe that it was structurally different from living cells (introduction figure I.19.1).

Bacterial viruses, known as **bacteriophages,** were first described by Twort (1915) and later by d'Herelle (1917). d'Herelle observed their filterable nature and their ability to form **plaques** on an agar plate seeded with a lawn of the host bacterium (see figure 21.2). Both Twort and d'Herelle worked with coliform bacteria isolated from the intestinal tract.

The agar medium provided a fast, easy way to recognize, identify, and quantify bacteriophage infections.

Viruses that attack mammalian cells also form structures analogous to plaques when cultivated on growth media able to support mammalian cell growth. Rather than plaques, the structures formed are described as **cytopathic effects (CPEs)** (introduction figure I.19.2). The CPEs observed are dependent upon the nature of both the host cell and the invading virus. The same can be said about the nature of bacterial plaque formation. Mammalian cells require complex growth media, including blood serum, as well as prolonged incubation with the virus before CPEs are visible. Therefore, we will work with an *E. coli* bacteriophage that can be obtained from a pure culture collection (Exercise 21).

Having made sundry efforts, from time to time, to discover, if 'twere possible, the cause of the hotness or power whereby pepper affects the tongue (more especially because we find that even though pepper hath lain a whole year in vinegar, it yet retaineth its pungency); I did now place anew about 1/3 ounce of whole pepper in water, and set it in my closet, with no other design than to soften the pepper, that I could the better study it. This pepper having lain about three weeks in the water, and on two several occasions snow-water having been added thereto, because the water had evaporated away; by chance observing this water on the 24th of April, 1676,

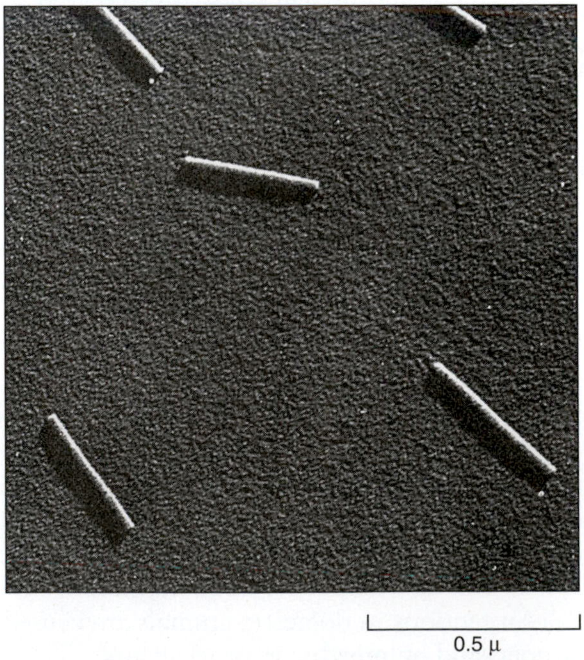

Figure I.19.2 Mammalian virus plaques showing different cytopathic effects (CPE). Photographed 5 days after infection of the growth medium (a single layer of monkey kidney cells) with the various mammalian viruses. When in monolayers, the viruses are able to form plaques (a form of CPE), which can be detected macroscopically.

From *Diagnostic Procedures for Viral. Rickettsial and Chlamydial Infections,* 5th Edition. Copyright © 1979 by the American Public Health Association. Reprinted with permission.

0.5 μ

Figure I.19.1 Tobacco mosaic virus. Electron micrograph (× approximately 70,000). Compare length and width with a rod-shaped bacterium.

© Omikron/Photo Researchers, Inc.

I saw therein, with great wonder, incredibly many very little animalcules, of divers sorts; and among others, some that were 3 or 4 times as long as broad, though their whole thickness was not, in my judgment, much thicker than one of the hairs wherewith the body of a louse is beset…. The second sort of animalcules consisted of a perfect oval. They had no less nimble a motion than the animalcules first described, but they were in much greater numbers. And there was also a third sort, which exceeded both the former sorts in number. These were little animals with tails, like those that I've said were in rainwater.

The fourth sort of little animals, which drifted among the three sorts aforesaid, were incredibly small; nay, so small, in my sight, that I judged that even if 100 of these very wee animals lay stretched out one against another, they could not reach to the length of a grain of coarse sand; and if this be true, then ten thousand of these living creatures could scarce equal the bulk of a coarse sand-grain.

I discovered yet a fifth sort, which had about the thickness of the last-said animalcules, but which were near twice as long.

DOBELL, *Antony van Leeuwenhoek and His Little Animals*

Definitions

Bacteriophage. A virus that infects bacteria. Frequently termed *phage*.

Cyst. A resting structure found in some protozoa.

Cytopathic effects (CPEs). Cellular changes caused by an infecting animal virus.

Giardiasis. Intestinal disease caused by *Giardia lamblia* found in contaminated drinking water. Sometimes called *beaver fever* because it can be spread by wild animals.

Mycology. The study of fungi.

EXERCISE 19

Getting Started

When you hear the word "fungus," what does it suggest? Mushrooms? **Toadstools?** Moldy fruit and yeast? All of these are examples of fungi. Fungi such as mushrooms and toadstools are macroscopic fungi, which can usually be identified without the aid of a microscope, whereas yeast and molds require microscopy for identification. However, because fungi are considerably larger than bacteria, they can be easier to identify.

Molds, yeasts, and perhaps another group, the **lichens,** are all members of the true fungi (Eumycota). The lichens are placed with the fungi for convenience because they represent dual-thallus plants composed of an alga and a fungus. The two fungus subgroups are the **nonfilamentous fungi** (yeasts), which are unicellular, and the **filamentous fungi** (molds), which are multicellular and have true filaments **(hyphae).** The hyphae are either nonseptate **(coenocytic)** or septate (figure 19.1). The nonseptate filaments are multinucleate, whereas the septate filaments contain either one or more nuclei per unit. This structural difference is important taxonomically in that one of the four classes of fungi, the Zygomycetes (table 19.1), is distinguished from the other classes by its lack of septate hyphae. Also, in contrast to the other three classes, it contains only a few human pathogens but numerous plant pathogens. Some authors divide the Zygomycetes into two classes, the Zygomycetes and the Oomycetes. The Zygomycetes are terrestrial fungi, and the Oomycetes are aquatic fungi containing the preponderance of plant pathogens. Fungus classification, although still somewhat in a state of flux, continues to improve with time.

Why all this interest in fungi? As with other forms of life, there are both the "good" and the "bad" fungi. Most are **saprophytes,** meaning that they decompose dead matter into a form that can be used to support all sorts of living matter. There are also the fungi that cause disease in plants or animals. For example, studies show that fungi unable to

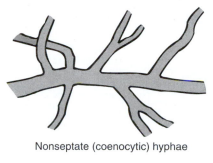

Nonseptate (coenocytic) hyphae

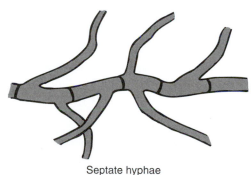

Septate hyphae

Figure 19.1 The two major hyphal types found in fungi.

synthesize their own nutrients invade a plant for these nutrients and thereby destroy it. For this purpose, they have spearlike hyphae adapted for invasive growth. Strangely enough, the same type of hyphae are also involved in the formation of various multicellular organs. The multicellular organs in turn can regenerate hyphae. Today, there is mounting interest worldwide concerning the impact of fungi on plant disease. Fungi are, in essence, **opportunists,** causing disease primarily only when an animal is in a weakened condition.

Examples illustrating the diverse morphology of yeasts and molds when examined with the microscope are shown in color plate 12 and figures 19.2, 19.3, 19.4, and 19.5. The beauty of fungus identification is that they can often be identified to the genus level simply by their macroscopic and microscopic growth characteristics when cultivated

Table 19.1   Classification of the Fungi

	Class			
	Zygomycetes	**Ascomycetes**	**Basidiomycetes**	**Deuteromycetes**
Mycelium	Nonseptate	Septate	Septate	Septate
Sexual spores	Oospore (not in a fruiting body) found in aquatic forms; Zygospore (not in a fruiting body) found in terrestrial forms	Ascospores, borne in an ascus, usually contained in a fruiting body	Basidiospores, borne on the outside of a clublike cell (the basidium), often in a fruiting body	None
Asexual spores	Zoospores, motile; Sporangiospores, nonmotile, contained in a sporangium	Conidiospores, nonmotile, formed on the tip of a specialized filament, the conidiophore	Same as Ascomycetes	Same as Ascomycetes
Common representatives	Downy mildews, potato blight, fly fungi, bread mold (*Rhizopus*)	Yeasts, morels, cup fungi, Dutch elm disease, ergot	Mushrooms, puff balls, toadstools, rusts, smuts, stinkhorns	Mostly imperfect Ascomycetes and some imperfect Basidiomycetes*

*Some of these fungi will no doubt form sexual spores in the right environment. In this event, they would need to be reclassified.

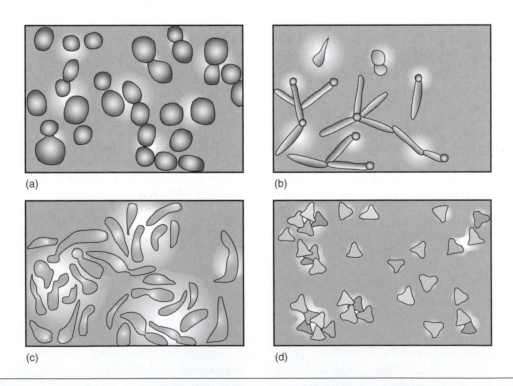

(a) (b) (c) (d)

Figure 19.2 Drawings of vegetative cell morphology of some representative yeasts. (*a*) *Saccharomyces cerevisiae*, round to oval cells; (*b*) *Candida sp.*, elongate to oval cells with buds that elongate, forming a false mycelium; (*c*) *Selenotila intestinalis*, lenticular cells; (*d*) *Trigonopsis variables*, triangular cells.

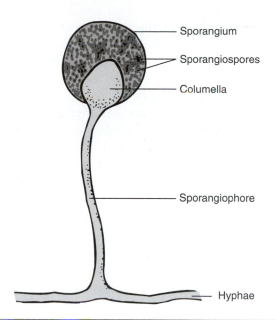

Figure 19.3 Intact asexual reproductive structure of the zygomycete *Rhizopus nigricans*. Note nonseptate (coenocytic) hyphae and sporangiophore.

Labels: Sporangium, Sporangiospores, Columella, Sporangiophore, Hyphae

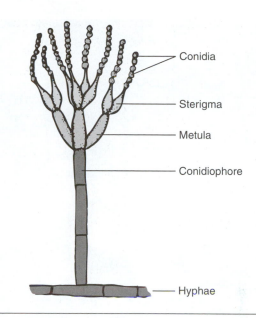

Figure 19.5 Intact asexual reproductive structure of the ascomycete *Penicillium*. Note the absence of a columella and a foot cell. Also note the symmetrical attachment of the metulae to the conidiophore—an important diagnostic feature for differentiation within the genus. Also note septate hyphae.

Labels: Conidia, Sterigma, Metula, Conidiophore, Hyphae

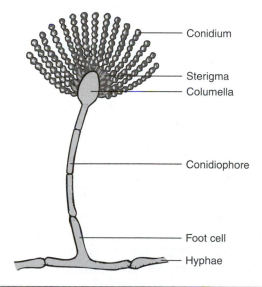

Figure 19.4 Intact asexual reproductive structure of the ascomycete *Aspergillus niger*. Note the presence of a foot cell, a columella, a septate conidiophore, and hyphae.

Labels: Conidium, Sterigma, Columella, Conidiophore, Foot cell, Hyphae

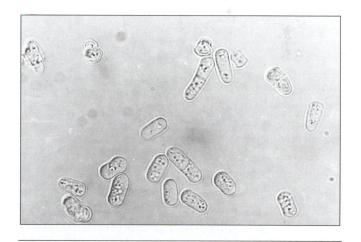

Figure 19.6 *Schizosaccharomyces pombe*, one of two yeasts known to multiply vegetatively by fission.
Courtesy of Dr. David Yarrow, The Central Bureau for Fungus Cultures, Baarn, Holland.

on various nutritional media. Most yeasts reproduce asexually by **budding.** An exception is the genus *Schizosaccharomyces*, which has the same vegetative reproduction process as bacteria—**fission** (figure 19.6). A microscopic study of cell and bud morphology can determine if a yeast is a member of the genus *Saccharomyces* or perhaps another genus with a different morphology, for example, *Selenotila* or *Trigonopsis* (figure 19.2). Fungi can be differentiated from bacteria by their larger size. Yeast and mold identification to the species level often requires additional morphological and physiological studies.

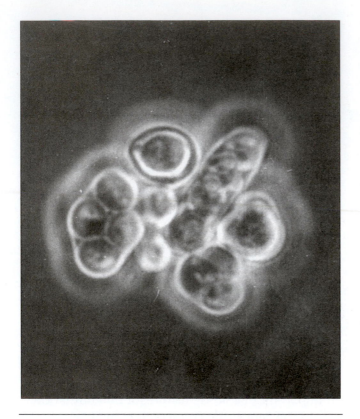

Figure 19.7 Sexual asci and ascospores of *Saccharharomyces cerevisiae*, showing asci with four or perhaps fewer spores.

Courtesy of the University of Washington Photo Library.

Morphologically, some yeasts form sexual spores **(ascospores)**, which are borne inside an **ascus.** Examples are *Saccharomyces cerevisiae* (figure 19.7) and *Schizosaccharomyces pombe*.

Some pathogenic yeasts exhibit a property known as **dimorphism** in which they can grow as yeasts or form mycelia. Dimorphism can be due to nutrients available or temperature. The genus *Candida* forms oval to elongate buds when grown on the surface of Sabouraud's dextrose agar (figure 19.2*b*), **chlamydospores** and **blastospores** when grown on cornmeal agar (figure 19.8), and **germ tubes** (figure 19.9) when grown in serum or egg albumin.

Physiological tests used for identifying yeast to the species level evaluate the yeast's ability to **assimilate** and/or **ferment** growth media containing various sugars as the sole carbohydrate source. Fermentation tests are conducted by inoculating tubes of broth containing different sugars with a drop of the test yeast. Each tube also contains a small inverted glass tube (Durham tube) to detect gas production.

Following incubation, the presence of gas (CO_2) in the tube constitutes a positive test for fermentation. The presence of yeast sediment or a change in color of the pH indicator is not, in itself, indicative of fermentation. The biochemical pathway for fermentation of sugar into alcohol, CO_2, and other end products, the Embden–Meyerhof pathway **(glycolytic pathway),** is discussed in Nester et al., p. 134.

The media used for initial isolation of many fungi is malt extract agar for the yeasts and Sabouraud's dextrose agar, with or without antibiotics to inhibit the growth of contaminant bacteria, for the filamentous fungi. The temperature of incubation depends on the organisms sought; 20°–25°C is suitable for most yeasts and filamentous fungi. One exception is *Aspergillus fumigatus*, which grows well at 45°C, a temperature that inhibits growth of most other fungi.

Many of the medically important fungi are found in the classes Ascomycetes and Deuteromycetes (see table 19.1). Most of their infections are opportunistic, limited to cutaneous or subcutaneous tissues. Such infections can sometimes progress to systemic involvement with the possibility of death. According to fungal expert Al-Doory, the use of new medical technologies, such as prolonged or extensive use of antibiotics, anticancer agents, and immunosuppressive drugs in organ transplants, is expected to continue, thus increasing the ever-present risk of opportunist fungal infections.

Clinically, there are three types of fungal diseases, or **mycoses:** dermatomycoses, subcutaneous mycoses, and systemic yeast and yeastlike infections.

1. **Dermatomycoses** are infections of the skin, hair, and nails caused by a group of filamentous fungi commonly called dermatophytes. They rarely invade subcutaneous tissues. They show rudimentary morphology, appearing only as mycelial growth on skin and nails or as fragments of **mycelium** and **arthrospores** arranged inside and outside of hair. In all instances, they form circular lesions described as **ringworm** (color plate 13), often caused by *Trichophyton* species (figure 19.10). However, in culture, they form filamentous colonies and asexual reproductive spores.

2. **Subcutaneous mycoses** are caused by either filamentous or **dimorphic** yeastlike fungi. They also remain fixed at the site of infection.

Table 19.2   Some Important Pathogenic Yeasts or Yeastlike (Dimorphic) Organisms

Organism	Morphology	Ecology & Epidemiology	Diseases	Treatment
Cryptococcus neoformans	Single budding cells, encapsulated	Found in soil and pigeons' nests. No transmission between humans and animals. May be opportunistic.	Meningitis, pneumonia, skin infections, visceral organs	Amphotericin B
Candida albicans	Budding cells, pseudomycelium formation chlamydospores	Normal inhabitants of mouth, intestinal tract. Opportunistic infections.	Thrush, vaginitis, nails, eyes, lungs, systemic infections	Alkaline mouth and douche washes; parahydroxy-benzoic acid esters; amphotericin B
Blastomyces dermatitidis	37°C: single large budding cells 20°C: mold with conidia	Disease of North America and Africa. Found occasionally in nature. No transmission between humans and animals.	Primarily lungs, also skin and bones	High-calorie, high-vitamin diet; bed rest; aromatic diamidines; amphotericin B
Paracoccidioides braziliensis (*Blastomyces braziliensis*)	37°C: single and multiple budding cells 20°C: mold with white aerial mycelium	Confined to South America. Workers in close association with farming.	Chronic granulomatous infection of mucous membranes of mouth, adjacent skin, lymph nodes, viscera	Sulfonamides; amphotericin B
Histoplasma capsulatum	37°C: single small budding cells 20°C: mold with tuberculate chlamydospores	Saprophyte in soil. No transmission between humans and animals. Epidemics from silos, chicken houses, caves, etc.	Primarily lungs, may spread to reticuloendothelial system	Amphotericin B
Coccidioides immitis	37°C: thick-walled, endospore-filled spherical cells 20°C: mold with arthrospores	Disease primarily of arid regions, such as San Joaquin Valley. Dust-borne disease. No transmission between humans and animals.	Primarily lungs, may disseminate particularly in African-Americans. Highly fatal	Bed rest; amphotericin B; surgery for lung lesions

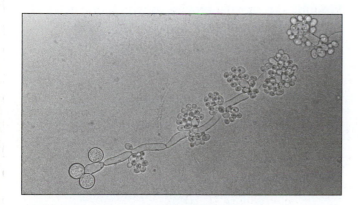

Figure 19.8 Chlamydospores and smaller blastospores (attached to pseudohyphae) of *Candida albicans* grown on cornmeal agar. Preparation stained with methylene blue.

Courtesy of the Upjohn Co.

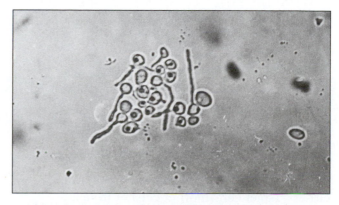

Figure 19.9 Germ tubes formed by *Candida albicans* grown on egg albumin. Phase-contrast magnification.

Courtesy of the Upjohn Co.

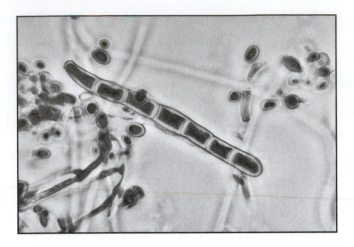

Figure 19.10 *Trichophyton rubrum*, a common dermatophytic fungus that causes ringworm.
Courtesy of CDC.

3. **Systemic yeast** and **yeastlike infections** originate in the lungs and can spread to other parts of the body. Examples include **histoplasmosis,** which is caused by the dimorphic yeastlike fungus *Histoplasma capsulatum*, and **coccidioidomycosis** (San Joaquin fever), caused by another yeastlike fungus, *Coccidioides immitis*. *Coccidioides immitis* infection is diagnosed in humans using the ELISA test (Exercise 28).

Definitions

Arthrospores. Thick-walled asexual spores formed by septate hyphae breaking apart.

Ascospore. Sexual spore characteristic of the fungus class Ascomycetes.

Ascus. Saclike structure containing ascospores.

Assimilation. Ability, in the presence of oxygen, to utilize carbohydrates for growth.

Blastospore. Asexual spore formed by budding from a cell or from hyphae.

Budding. An asexual process of reproduction in which a daughter cell (bud) evolves from either a larger cell (mother cell) or from a hypha.

Chlamydospore. A resistant hyphal cell with a thick wall; it eventually separates from the hyphae and functions as a spore.

Coccidioidomycosis. An upper respiratory tract infection caused by the dimorphic yeastlike organism *Coccidioides immitis*.

Coenocytic. A multinuclear mass of protoplasm resulting from repeated nuclear division unaccompanied by cell division.

Columella. A swelling of the sporangiophore at the base of the sporangium that acts as a support structure for the sporangium and its contents.

Conidia (singular, *conidium*). Asexual spores produced from either the tip or side of the conidiophore.

Conidiophores. Specialized hyphae that bear asexual spores.

Conidiospores. Asexual spores.

Culture spherule. A thick-walled, sphere-shaped cell containing many small endospores, characteristic of the tissue phase of *Coccidioides immitis*.

Cyst. A resting (dormant) spore.

Dermatomycosis. A disease of the skin caused by infection with a fungus.

Dimorphic. Ability to exist in two forms, e.g., in the fungi, either a mycelial form or a yeastlike form.

Fermentation. Growth, in the absence of oxygen, in which the final electron acceptor is an organic compound.

Filamentous fungi. Fungi that grow by hyphae rather than budding.

Fission. An asexual process in which one cell splits into two or more daughter cells.

Foot cell. A cell located at the base of the conidiophore in the genus *Aspergillus*.

Germ tube. A tubelike outgrowth from an asexual yeast cell that develops into a hypha.

Glycolytic pathway. An initial series of fermentation steps in which carbohydrates are degraded. Often called the *Embden–Meyerhof pathway*.

Histoplasmosis. A pulmonary infection caused by *Histoplasma capsulatum*, a dimorphic fungus.

Hypha (plural, *hyphae*). Threadlike fungal filament that forms a mycelium.

Lichen. A symbiotic relationship between a fungus and an alga. See Nester et al. for additional information.

Metula. A branch (or branches) at the tip of the conidiophore that supports sterigmata.

Mold. A filamentous fungus often appearing as woolly growth on decaying materials.

Mycelium. A fungal mat made of tangled hyphae.

Mycology. The study of fungi.

Mycoses. Diseases caused by fungi.

Nonfilamentous fungi. Fungi devoid of hyphae, e.g., yeast.

Opportunist. An organism capable of causing disease only when host defense mechanisms are impaired.

Rhizoids. Rootlike structures made of fungus hyphae that are able to penetrate various substrates in order to anchor the fungus so that it can obtain nutrients.

Ringworm. Contagious fungal disease of the hair, skin, or nails. See color plate 13.

Saprophyte. An organism that obtains nourishment from decayed organic matter.

Spherule. A large, thick-walled structure filled with fungal endospores.

Sporangiospore. Asexual reproductive spore found in the Zygomycetes.

Sporangium. The sac containing asexual spores.

Sterigma (plural, *sterigmata*). A specialized hypha that supports either a conidiospore or a basidiospore.

Stolon. A runner, such as found in strawberry plants, made of horizontal hyphae from which sporangiospores and rhizoids originate. Stolons are characteristic of the class Zygomycetes.

Subcutaneous mycoses. Fungal infections that do not spread through the body but remain localized.

Systemic yeast. Yeast found in various parts of the body.

Toadstool. A large, filamentous, fleshy fungus with an umbrella-shaped cap.

Yeast. A nonfilamentous fungus often found in nature on fermenting fruits and grains.

Yeast dimorphism. Existing in two growth forms, such as the mold phase (hyphal filaments) and yeast phase (single cells) of pathogenic fungi.

Yeastlike infections. Diseases such as histoplasmosis and coccidioidomycosis caused by dimorphic fungi.

Objectives

1. Describe what fungi are and how to distinguish them from one another, including how to distinguish members of the two major groups of fungi (nonfilamentous and filamentous).

2. Explain how culturing yeasts and fungi is different from culturing bacteria.

3. Interpret macroscopic and microscopic colonial and vegetative cell morphology in order to identify *Saccharomyces*, *Rhizopus*, *Penicillium*, and *Aspergillus*.

Prelab Questions

1. What is the difference in shape between the asexual structures of *Rhizopus* and *Penicillium*?

2. Which fungal class does not contain septa?

3. Name a disease *Candida albicans* can cause.

Materials

Per team of 2-4 students

Cultures of the following:

Sabouraud's dextrose broth cultures (48 hr, 25°C) of *Saccharomyces cerevisiae*

Sabouraud's dextrose agar petri dish cultures (3–5 days, 25°C) of *Rhizopus nigricans*, *Aspergillus niger*, and *Penicillium notatum*

Glucose broth tube containing a Durham tube, 1

Maltose broth tube containing a Durham tube, 1

Lactose broth tube containing a Durham tube, 1

Glucose acetate yeast sporulation agar, 1 plate

Sabouraud's dextrose agar, 1 plate

Sterile droppers, 4

Dissecting microscope(s)

Ruler divided in mm

Dropping bottle containing methylene blue

Slides for wet mounts

First Session

Suspend the broth culture of *Saccharomyces cerevisiae*.

1. Yeast fermentation study. Inoculate each of the carbohydrate fermentation tubes (glucose, lactose, and maltose) with a loopful of *Saccharomyces cerevisiae*. Place the properly labeled tubes in a container, and incubate at 25°–30°C for 48 hours.

2. Yeast colonial and vegetative cell morphology study. Label a Sabouraud's dextrose agar plate "*S. cerevisiae*." With a sterile dropper, inoculate the agar surface with a *small* drop of yeast. Allow the inoculum to soak into the agar before incubating right side up in the 25°–30°C incubator for 48 hours.

3. Yeast sexual sporulation study. With a sterile dropper, inoculate the center of the sporulation agar plate with a small drop of the *S. cerevisiae* broth culture. Allow the inoculum to soak into the agar before incubating right side up in the 25°–30°C incubator for 48 or more hours. Cultures freshly isolated from nature generally sporulate much faster than laboratory-held cultures.

4. Colonial characteristics of petri dish cultures of *Rhizopus nigricans*, *Aspergillus niger*, and *Penicillium notatum* or other species of *Penicillium*.

   Visually examine each petri dish culture noting the following:

   a. Colony size. With a ruler, measure the diameter in mm.

   b. Colony color. Examine both the upper and lower surfaces.

   c. Presence of soluble pigments in the agar medium.

   d. Colony texture (such as cottony, powdery, or woolly).

   e. Colony edge (margin). Is it regular or irregular?

   f. Colony convolutions (ridges). Are they present?

   Enter your findings in table 19.5 of the Laboratory Report.

5. Morphological study of asexual fruiting structures found in *Rhizopus*, *Aspergillus*, and *Penicillium* species.

   The best way to make studies of this type is with a covered slide culture. In the event none is available, you can attempt to do so with your petri dish culture. A major problem is the density of growth in the petri dish culture, which makes it difficult to find intact asexual reproductive structures. However, they can often be found for *R. nigricans*, in that, like strawberries, it has **stolons,** which enable it to spread and attach to the underside of the petri dish lid. A dissecting microscope is preferred for making initial observations. If not available, the light microscope's low-power objective lens can also be used. The low-power objective lens more than doubles the magnification obtained with the dissecting microscope.

   Procedurally do as follows: Place the covered petri dish culture on the stage of either the dissecting microscope or the light microscope, and examine a sparsely populated area of the colony for the presence of asexual reproductive structures (figure relating to the fungus culture being examined, 19.3, 19.4, or 19.5).

   Note: Never smell fungus cultures—spore inhalation may cause infection. When you first observe fruiting bodies, stop moving the petri dish. Keeping the air currents to a minimum to avoid spore dispersal, carefully remove the petri dish cover and re-examine to determine whether you can in fact see the various parts of the fruiting body as described in the figure for that fungus. In the case of *R. nigricans*, you should be able to see fruiting bodies by examining the underside of the petri dish lid. It may take as long as 4–5 days' incubation before finding *Rhizopus* fruiting bodies with stolons. If intact fruiting bodies are found for any of the three filamentous fungi, make drawings of their asexual reproductive structures in part 5 of the Laboratory Report. Label the parts in a manner similar to that used in figures 19.3, 19.4, and 19.5. Covered slide cultures are the answer if fruiting bodies cannot be found for the *A. niger* and *P. notatum*.

6. Detailed examination of **sporangiospores, conidia,** and if present, chlamydospores.

For observing sporangiospores (*R. nigricans*, color plate 12) and conidia (*A. niger* and *P. notatum*), prepare a wet mount by adding a drop of methylene blue to a slide. Remove some aerial growth with a loop and add it to the methylene blue. Add a coverslip. Observe the slide with the low-power and high-dry objective lenses of the microscope.

Chlamydospores can be found in both surface and submerged *R. nigricans* mycelium. They are elongated, brown in color and have thick walls. Prepare a wet mount like above and observe with both the low-power and high-dry objective lenses.

Note: You may wish to first search for spores in both the inner and outer fringes of the colony using the low-power objective lens. To better search the inner colony surface, make the area less dense by first removing some of the aerial growth with a loop. Flame the loop to destroy the spores. In part 6 of the Laboratory Report, prepare and label drawings of the various asexual spore types found.

Some morphological characteristics of value for identification include the following for the three cultures under examination.

1. ***Rhizopus nigricans***
   a. Has nonseptate (coenocytic) hyphae.
   b. Contains **rhizoids.** See the underside of the petri dish of an older culture.
   c. Details of fruiting body (figure 19.3): Note the nonseptate stem (sporangiophore), the swelling at the tip of the sporangiophore **(columella),** and the sac that encloses the columella **(sporangium),** which contains the asexual reproductive spores (sporangiospores).

2. and 3. *Aspergillus niger* and *Penicillium notatum*

These molds have fruiting bodies somewhat similar in appearance. Both have brush-like structures comprising bottle-shaped cells **(sterigmata)** to which are attached long chains of asexual reproductive spores **(conidiospores).** They differ in that the genus *Aspergillus* has a swollen cell at the base of the stem **(conidiophore)** known as the **foot cell,** as well as a columella to which the sterigmata are attached (figure 19.4). The sterigmata may oc-

cur in one or two series, depending upon the species involved. Finally, in the genus *Penicillium*, the branching at the tip of the conidiophore can be either symmetrical (figure 19.5) or asymmetrical, depending on where the **metulae** are attached to the conidiophore. When all of the metulae are attached at the tip of the conidiophore, branching of the sterigmata and attached conidia will appear symmetrical (figure 19.5). If one of the metulae in figure 19.5 is attached below the tip of the conidiophore, an asymmetrical branching occurs. This is an important diagnostic feature for differentiation within the genus *Penicillium*.

Second Session

1. Yeast fermentation study. Examine the fermentation tubes for the following and record your results in table 19.3 of the Laboratory Report:
   a. Presence or absence (+ or −) of cloudy broth (growth).
   b. Presence or absence (+ or −) of gas in the inverted Durham tube.
   c. Change in color of the pH indicator dye. A change to a yellow color is indicative of acid production.

Note: Gas production is indicative of fermentation (glycolysis). To detect false-negative results caused by super saturation of the broth, all tubes giving an acid reaction should be shaken lightly and the cap vented. This operation is frequently followed by a rapid release of gas. All positive fermentation reactions with a carbohydrate sugar are accompanied by positive assimilation of that carbohydrate, evidenced by increased clouding of the broth; however, sugars may be assimilated without being fermented.

2. Yeast colonial and vegetative cell morphology study.
   a. Colony characteristics. If possible, observe the Sabouraud's agar plate over a 5- to 7-day incubation period. Make note of the following in table 19.4 of the Laboratory Report: colony color; colony consistency (soft, firm—probe the colony with a sterile needle for this determination); colony diameter (mm); colony surface (rough or

smooth, flat or raised); and appearance of the colony edge (circular or indented).

b. Vegetative cell morphology. Remove a loopful of surface growth from each colony and prepare wet mounts as above. Observe with the high-dry objective lens, noting the shape and size of the cells and the presence or absence of pseudohyphae (figure 19.2b). Prepare and label a drawing of the yeast in part 2b of the Laboratory Report.

3. Sexual sporulation study. With a sterile loop, touch the S. cerevisiae colony on the glucose–acetate sporulation agar plate and prepare a wet mount as above. Observe with the high-dry objective lens, looking for the presence of asci containing one to four or perhaps more ascospores (figure 19.7). Prepare and label drawings of your findings in part 3 of the Laboratory Report.

Note: In the event you do not find asci, reincubate the plate up to one week, perhaps even longer, and re-examine periodically. Some yeast strains take longer than others to produce sexual spores.

4., 5., and 6.   Filamentous fungi. Complete any remaining morphological studies.

Note: If an ocular micrometer is available and time permits, you may wish to make measurements of some of the various morphological structures, perhaps a comparison of asexual spore sizes of different filamentous fungi for example. Appendix 5 contains information on use, calibration, and care of the ocular micrometer.

EXERCISE

19

Laboratory Report:
Microscopic Identification of Fungi

Results (Nonfilamentous Fungi)

1. Fermentation Study. Examine tubes and record results (+ or −) in table 19.3. For details, see Procedure, session 2, step 1, p. 147.

Table 19.3   *Saccharomyces cerevisiae* Fermentation Activity in Tubes of Broth Containing Different Carbohydrate Sugars

Yeast	Glucose			Maltose			Lactose		
	Cloudy	Gas	Acid	Cloudy	Gas	Acid	Cloudy	Gas	Acid
Saccharomyces cerevisiae									

2. Yeast Colonial and Vegetative Cell Morphology Study.

   a. Colony characteristics. For details, see Procedure, session 2, step 2a, p. 147. Enter results in table 19.4.

Table 19.4   *Saccharomyces cerevisiae* Colonial Characteristics on Sabouraud's Dextrose Agar Plates

Yeast Strain	Colonial Morphology				
	Colony Color	Consistency	Diameter (mm)	Surface Appearance	Edge Appearance
Saccharomyces cerevisiae					

b. Vegetative cell morphology. For details see Procedure, session 2, step *2b*, p. 147–148. Enter results below.

Saccharomyces cerevisiae

3. Sexual Sporulation Study *(Saccharomyces cerevisiae)*. Drawings of asci and ascospores. For details, see Procedure, session 2, step 3, p. 148.

4. Colonial characteristics of *Rhizopus*, *Aspergillus*, and *Penicillium* when grown on Sabouraud's dextrose agar. Describe in table 19.5.

Table 19.5   Colonial Characteristics of Three Filamentous Fungi Cultured for _____ Days on Sabouraud's Dextrose Agar

	Rhizopus	*Aspergillus*	*Penicillium*
Colony color			
Colony diameter (mm)			
Colony texture			
Colony convolutions			
Colony margin			
Soluble pigments in agar			

5. Drawings of their asexual reproductive structures (please label parts):

   Rhizopus   *Aspergillus*   *Penicillium*

6. Drawings of their asexual spores (please label parts):

   Rhizopus   *Aspergillus*   *Penicillium*

Questions

1. Explain the physiological differences between yeast fermentation and yeast assimilation of glucose.

2. Why is a loop rather than a pipet used to inoculate the sugar fermentation tubes?

3. Why would the growth of a pellicle, or film, on the surface of a broth growth medium be advantageous to the physiology and viability of that yeast?

4. What are some ways in which you might be able to differentiate *Rhizopus nigricans* from *Aspergillus niger* simply by visually observing a petri dish culture?

5. How can you determine whether or not a green, woolly looking colony is *Aspergillus* or *Penicillium?*

6. What problems might you have in identifying a pathogenic fungus observed in a blood specimen? What might you do to correct such problems?

7. In what ways can we readily distinguish
   a. fungi from algae?

   b. fungi from bacteria?

   c. fungi from actinomycetes?

8. Define *opportunistic fungus*. Provide some examples. Are all medically important fungi opportunistic? Discuss your answers.

9. Name three pathogenic fungi that exhibit dimorphism. Describe the type of dimorphism each exhibits and the laboratory conditions necessary to elicit it.

EXERCISE

20

Parasitology: Protozoa and Helminths

Getting Started

Because the natural histories of parasitic diseases differ in some important respects from those of bacterial diseases, they merit a separate introductory laboratory experience with parasites, the diseases they cause, and techniques used to diagnose them.

The distinguishing features of parasitic life are the close contact of the parasite with the host in or on which it lives and its dependency on the host for life itself. This special association has led to the evolution of three types of adaptations not found in the free-living relatives of the parasites: loss of competency, special structures, and ecological ingenuity.

Parasites have become so dependent on their hosts for food and habitat that they now experience a *loss of competency* to live independently. They usually require a specific host, and many have lost their sensory and digestive functions; these are no longer important for their survival.

On the other hand, they have developed *special structures* and functions not possessed by their free-living relatives that promote survival within the host, such as hooklets and suckers for attachment. Parasites also have an increased reproductive capacity to compensate for the uncertainty in finding a new host. Tapeworms can produce up to 100,000 eggs per day.

Ecological ingenuity is demonstrated in the fascinating variety of infecting and transmitting mechanisms. This has led to very complex life cycles, which contrast markedly with the relatively simple lifestyles of their free-living counterparts. Parasites possess diverse life cycles, ranging from species that pass part of each generation in the free-living state, to others that require at least three different hosts to complete the life cycle. Some are simply transmitted by insects from one human host to a new host, or the insect may act as a host as well. Many protozoa develop resistant cysts, which enable them to survive in unfavorable environments until they find a new host. The eggs of flatworms and roundworms also have a protective coat.

These three strategies promote species survival and expansion by providing greater opportunities for finding and infecting new hosts, which is a continual problem for parasites. Successful interruption of these cycles to prevent their completion is an important feature of public health measures used to control diseases caused by parasites.

This exercise is designed to give you some practical experience with representative protozoan and helminthic parasites and with clinical methods used in their diagnosis and control. The following classification of parasites will serve as a guide to the examples you will be studying in this exercise. It is not a complete listing.

Protozoa

Protozoa, a subkingdom of the kingdom Protista, are unicellular eukaryotic organisms. They usually reproduce by cell division and are classified mainly according to their means of locomotion. Only one phyla, the Suctoria, which is closely related to the Ciliophora, does not contain animal pathogens. The remaining are classified as follows.

Sarcodina

Members of this subphylum move and feed slowly by forming cytoplasmic projections known as **pseudopodia** (false feet). They also form both **trophozoites** (vegetative form) and **cysts** (resistant, resting cells). Parasitic members include the **amoeba** *Entamoeba histolytica*, which causes amoebic dysentery in animals and humans (color plate 14). It ingests red blood cells and forms a four-nucleate cyst. Other amoeba species found in humans, such as *Entamoeba gingivalis*, are relatively harmless **commensals.**

Ciliophora

Members of this phylum have many short, hairlike cilia on their body surfaces that beat rhythmically by bending to one side. They contain two nuclei: a macronucleus and a micronucleus. Ciliophora is typified by the genera *Paramecium* (color plate 15) and *Stentor* (figure 20.1), both nonpathogens readily found in pond water. Another member, *Balantidium coli* (figure 20.2), is a common parasite in swine. It possesses both a cyst and trophozoite form and can infect humans, causing serious results.

Mastigophora

These protozoans propel themselves with one or more long, whiplike flagella. Some have more than one nucleus, and they usually produce cysts. Different species cause infections in the intestines, vagina, blood, and tissues. *Giardia lamblia* (color plate 16, figure 29.3) causes a mild to severe diarrheal infection. *Trichomonas vaginalis* (figure 20.3) is found in the urogenital region, where it causes a mild vaginitis in women. *Trypanosoma gambiense* (color plate 17) infects the blood via tsetse fly bites, causing trypanosomiasis, or African sleeping sickness, in cattle and humans. Cattle and other ungulates serve as reservoirs for this organism.

Sporozoa

Sporozoa are obligate, nonmotile parasites with alternating stages: The sexual reproductive stage is passed in the definitive insect host, and the asexual phase is passed in the intermediate human or animal host. The genus *Plasmodium* (color plates 18 and 19) includes the malarial species (*malariae, ovale, vivax, falciparum*), in which the **definitive host** is the female *Anopheles* mosquito and the **intermediate host** is humans. The genus *Coccidia* includes important intestinal parasites affecting fowl, cats, dogs, swine, sheep, and cattle. *Toxoplasma gondii* is a cat parasite that can harm the human fetus in an infected pregnant woman.

Helminths (Worms)

Helminths are multicellular eukaryotic organisms. Two of the phyla, Platyhelminthes (flatworms) and Nematoda (roundworms), contain pathogenic worms.

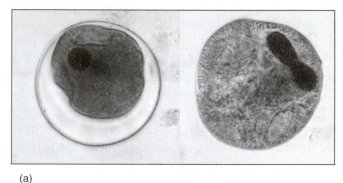

(a)

(b)

Figure 20.2 (*a*) *Balantidium coli* cysts viewed on 40×. The macronucleus can be seen on the photo on the right. (*b*) *Balantidium coli* trophozoite viewed on 40×, showing the nucleus overlaid on the macronucleus.

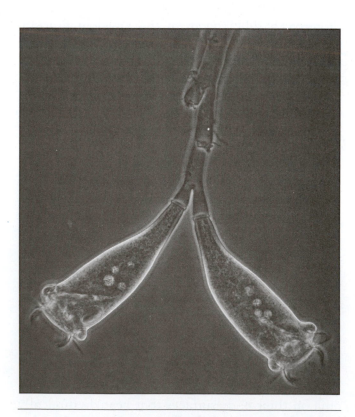

Figure 20.1 A *Stentor* viewed at 40× in pond water. Courtesy of Anna Oller, University of Central Missouri.

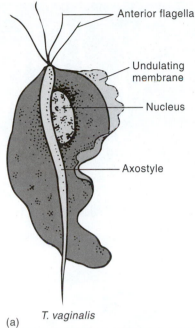

Anterior flagella

Undulating membrane

Nucleus

Axostyle

T. vaginalis

(a)

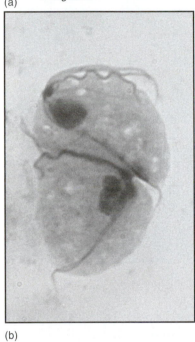

(b)

Figure 20.3 *(a) Trichomonas vaginalis.* Illustration of a typical mastigophoran protozoan. *(b) Trichomonas vaginalis* viewed on 100×. The flagella, nucleus, axostyle, and undulating membrane can all be seen at the top.

(b) Courtesy of Anna Oller, University of Central Missouri.

Phylum Platyhelminthes

Members of Platyhelminthes are flat, elongated, legless worms that are **acoelomate** and exhibit bilateral symmetry. This phylum contains three classes.

Class Turbellaria

Members of this class are free-living **planarians** (flatworms), such as are found in the genera *Dugesia* (figure 20.4) and *Planaria*.

Class Trematoda (Flukes)

Flukes have an unsegmented body, and many have suckers to hold them onto the host's intestinal wall. Many flukes have complex life cycles that require aquatic animal hosts. The *Schistosoma* species (figure 20.5)

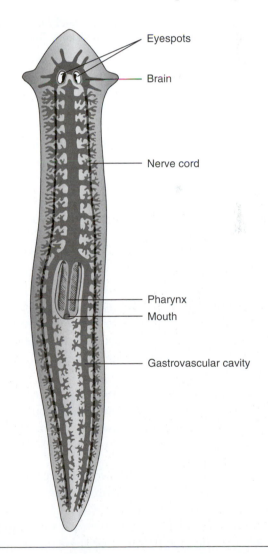

Eyespots

Brain

Nerve cord

Pharynx

Mouth

Gastrovascular cavity

Figure 20.4 Labeled line drawing of the genus *Dugesia*, a free-living planarian in the class Tubellaria.

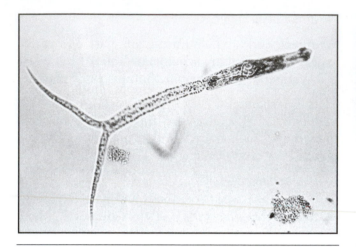

Figure 20.5 Schistosomal cercaria.

CDC/Minnesota Department of Health, R.N. Barr Library; Librarians Melissa Rethlefsen and Marie Jones, Prof. William A. Riley.

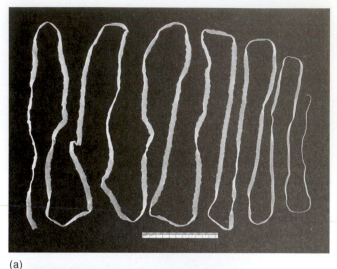

(a)

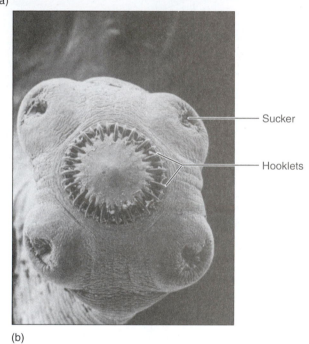

Sucker

Hooklets

(b)

Figure 20.6 (a) *Taenia saginata* adult tapeworm. (b) A tapeworm scolex showing both hooklets and suckers for attachment to the intestine. *Taenia saginata* (beef tapeworm) is essentially without hooklets, whereas *Taenia solium* (pork tapeworm) has both.

(a) Courtesy of CDC. (b) Stanley Flegler/Visuals Unlimited.

are bisexual trematodes that cause serious human disease. They require polluted water, snails, and contact with human skin for completion of their life cycles. *Clonorchis sinensis* and *Fasciola hepatica* (color plate 20) species are liver flukes acquired by eating infected raw fish and contaminated vegetables, respectively.

Class Cestoda (Tapeworms)

Tapeworms are long, segmented worms (figure 20.6a) with a small head (**scolex**) equipped with suckers and often hooklets (figure 20.6b, color plate 21) for attachment to the host's intestinal wall. The series of segments, or **proglottids,** contain the reproductive organs and thousands of eggs. These segments break off and are eliminated in the feces, leaving the attached scolex to produce more proglottids with more eggs. Figure 20.7 illustrates the life cycle of the tapeworm in a human. The symptoms of *Taenia* tapeworm infection are usually not serious, causing only mild intestinal symptoms and loss of nutrition—not so for the *Echinococcus* tapeworm, which causes a serious disease. All tapeworm diseases are transmitted by animals.

Phylum Nematoda

Members of Nematoda (roundworms) are present in large numbers in diverse environments, including soil, freshwater, and seawater. This phylum contains many agents of animal, plant, and human parasitic diseases. In contrast to the Platyhelminthes, these round, unsegmented worms are **coelomate** (have a body cavity) and have a complete digestive tract and separate sexes. Most require only one host and can pass part of their life cycle as free-living larvae in the soil. *Trichinella spiralis* (color

plate 22) requires alternate vertebrate hosts. Humans become infected when they ingest inadequately cooked meat, such as pork or bear containing the larval forms in the muscles. *Ascaris lumbricoides* (figure 20.8, a–b) is probably the most common worldwide of all the human helminths. Larva migrate through the lungs and up the

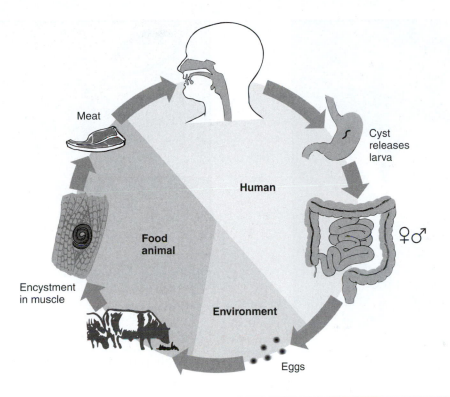

Meat

Human

Cyst releases larva

Food animal

♀♂

Encystment in muscle

Environment

Eggs

Figure 20.7 Life cycle of *Taenia saginata*. The adult tapeworm with scolex and proglottids is conceived from larvae in the human intestine.

trachea, where they are then swallowed and mature in the intestines. *Enterobius vermicularis* causes pinworm (figure 20.9), a very common condition in children characterized by anal itching. *Dirofilaria immitis* (color plate 23) is transmitted via mosquitoes and causes heartworms in animals. Left untreated, or treated with high doses of Ivermectin, animals will perish. Animals must be treated with low doses of Ivermectin over a period of time because a large number of dead worms can cause a lethal hypersensitivity or embolus.

Definitions

Acoelomate. Without a true body cavity. Typical of members of the Phylum Platyhelminthes (flatworms).
Amoeba. Unicellular organisms with an indefinite changeable form.
Cercaria. The last miracidium stage in which the larvae possess a tail.
Coelomate. With a true body cavity. Typical of members of the Phylum Nematoda (roundworms).
Commensal. A relationship between two organisms in which one partner benefits from the association and the other is unaffected.

Cysts. Dormant, thick-walled vegetative cells.
Definitive host. The host in which the sexual reproduction of a parasite takes place.
Intermediate host. The host that is normally used by a parasite in the course of its life cycle and in which it multiplies asexually but not sexually.
Merozoites. Schizont nuclei that become surrounded by cytoplasm and bud off as daughter cells or merozoites.
Miracidium. A free-swimming ciliate larva that seeks out and penetrates a suitable intermediate snail host, in which it develops into a sporocyst.
Planarian. Any flatworm of the genus *Planaria*.
Proglottid. Any of the segments of a tapeworm formed in the neck region by a process of strobilation (transverse fission).
Pseudopodia. Extensions of cytoplasm that aid in engulfing particles and functioning in motility of amoeboid cells.
Schizont. A stage in the life cycle of Sporozoa in which the nucleus undergoes repeated nuclear division without corresponding cell divisions.

(a)

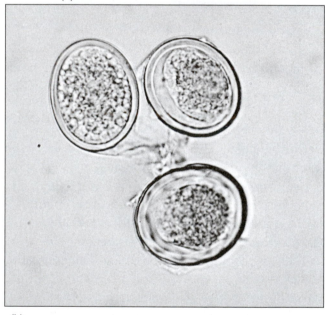

(b)

Figure 20.8 (*a*) *Ascaris lumbricoides* worms.
(*b*) *Ascaris lumbricoides* eggs.

(*a*) Courtesy of CDC. (*b*) © Anna Oller, University of Central Missouri.

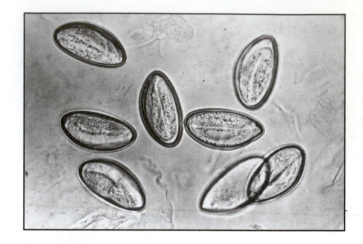

Figure 20.9 *Enterobius vermicularis* eggs.
Courtesy of CDC.

Objectives

1. Differentiate nonparasitic protozoans (*Amoeba proteus* and a *Paramecium* species) and a free-living planarian (flatworm) from parasitic protozoans like *Giardia lamblia*.
2. Describe the morphology of some free-living, trophozoite, and cystic forms of intestinal parasites using prepared slides, and malarial and trypanosome parasites using stained blood smears.
3. Explain the natural history and life cycle of an important human parasitic disease, schistosomiasis, using stained slides and a life-cycle diagram.

Prelab Questions

1. What is the common name for a trematode?
2. Provide the scientific name of a Mastogophoran.
3. What are the main steps in the life cycle of *Schistosoma mansoni*?

Scolex. The head of a tapeworm, which is used for attaching to the host's intestinal wall.

Sporocyst. A stage in the life cycle of certain protozoa in which two or more of the parasites are enclosed within a common wall.

Trophozoites. Vegetative forms of some protozoans.

Materials

Cultures of the following:

Living cultures of a *Paramecium* species, *Amoeba proteus*, and a *Dugesia* or *Planaria* species

If available, a fresh sample of quiescent, stagnant pond water, which often contains members of the above genera. Students may wish to bring their own pond water.

Planaslo solution, 1 or more dropping bottles

Depression slides (hanging drop slides)

The following commercially prepared slides:

Subkingdom Protozoa

Sarcodina (pseudopodia)

Entamoeba histolytica trophozoite and cyst stages

Ciliophora (cilia)

Paramecium trophozoite or *Stentor*

Balantidium coli

Mastigophora (flagella)

Giardia lamblia trophozoite and cyst stages

Trichomonas vaginalis

Trypanosoma gambiense

Sporozoa (nonmotile)

Plasmodium vivax or *P. falciparum* ring, amoeboid schizont stages

Subkingdom Helminths (worms)

Phylum Platyhelminthes (flatworms)

Class Turbellaria (free living)

Dugesia or *Planaria* species

Class Trematoda (flukes)

Schistosoma mansoni

   adult male

   adult female

   ovum (egg)

   ciliated miracidium

   infective ciliate cercaria

   sporocyst stage in snail liver tissue

Clonorchis sinensis or *Fasciola hepatica*

Class Cestoda (tapeworms)

Taenia solium trophozoite or *T. saginata*

Class Nematoda

Ascaris lumbricoides

Dirofilaria immitis

PROCEDURE

Your study will consist of these procedures:

1. Observe the movements and structure of some living nonparasitic protozoans and worms often found in pond water.

2. Examine commercially prepared stained blood and fecal slides that contain human protozoan parasites.

3. Microscopically compare the structures of parasitic worms to those of their free-living relatives.

4. Study the natural history and life cycle of the human parasitic disease schistosomiasis.

Note: If the number of prepared slides is limited, these procedures may be performed in a different order to facilitate sharing.

1. Examination of free-living cultures.

   a. Pond water examination. Prepare a wet mount (Exercise 3) using petroleum jelly to adhere a coverslip to a glass slide. Examine initially with the low-power objective lens and later with the high-dry objective lenses. Observe the mode of locomotion of any amoeboid or paramecium-like protozoans found. If their movements are too rapid, add a drop of Planaslo to slow them down. Describe their movements and prepare drawings in part 1*a* of the Laboratory Report.

   b. Examination of fresh samples of an amoeba (such as *Amoeba proteus*), paramecium (for example, *Paramecium caudatum*), and a free-living flatworm (such as *Dugesia* or *Planaria* species). Use wet mount slide preparations and examine as described in 1*a* for pond water. Record your observations in part 1*b* of the Laboratory Report.

2. Examination of stained slides for trophozoites and cysts in feces.

   a. Using the oil immersion objective lens, examine prepared slides of a protozoan, either the amoeba *Entamoeba histolytica* or the flagellate *Giardia lamblia*. In the trophozoite stage, observe the size, shape, number of nuclei, and presence of flagella or pseudopodia. In the cyst stage, look for an increased number of nuclei and the thickened cyst wall.

    b. Sketch an example of each stage, label, and record in part 2 of the Laboratory Report.

3. Examination of protozoans present in stained blood slides.

    a. Use the oil immersion objective lens to explore a smear of blood infected with *Plasmodium vivax*, and locate blood cells containing the parasite. After a mosquito bite, the parasites are carried to the liver, where they develop into **merozoites**. Later, they penetrate into the blood and invade the red blood cells, where they go through several stages of development. The stages are the delicate ring stage (color plate 18), the mature amoeboid form, and the **schizont** stage (color plate 19), in which the organism has divided into many individual infective segments that will then cause the red blood cell to rupture, releasing the parasites, which can then infect other cells. Sketch the red blood cells with the infective organism inside them, and note any changes in the red blood cell shape, pigmentation, or the presence of granules due to the effect of the parasite. Identify and label the stage or stages seen, as well as the species, in part 3a of the Laboratory Report.

    b. Examine the trypanosome blood smear (color plate 17) with the oil immersion lens, and locate the slender flagellates between the red blood cells, noting the flagellum and undulating membrane. Sketch a few red blood cells along with a flagellate in part 3b of the Laboratory Report.

4. Comparison of a free-living worm with its parasitic relative.

    a. Observe a prepared slide of a free-living flatworm (*Dugesia* or *Planaria* species) with the low-power objective lens. Note the pharynx, digestive system, sensory lobes in the head region, and the eyespots

(see figure 20.4). Next examine a parasitic fluke such as *Clonorchis* or *Fasciola*. Note the internal structure, especially the reproductive system and eggs if female, and the organs of attachment, such as hooklets or round suckers.

    b. Sketch each organism in part 4a of the Laboratory Report, and label the main features of each. Describe the main differences between the fluke and the free-living planaria.

    c. Examine prepared slides of a tapeworm (*Taenia* species, figure 20.6), observing the small head, or scolex, and the attachment organs—the hooklets or suckers. Then locate the maturing proglottids along the worm's length. The smaller proglottids may show the sex organs better; a fully developed proglottid shows the enlarged uterus filled with eggs. Sketch, label, and describe its special adaptations to parasitic life in part 4b of the Laboratory Report.

    d. Observe a prepared slide of *Ascaris lumbricoides*, noticing differences between the flatworm, fluke, and tapeworm. Sketch in part 4c of the Laboratory Report.

    e. Observe a prepared slide of *Dirofilaria immitis* on 40×, noting the differences between the other classes of worms. Sketch in part 4d of the Laboratory Report.

5. Life cyle of *Schistosoma mansoni* and its importance in the control of schistosomiasis.

    a. Assemble five or six slides showing the various stages in the schistosoma life cycle: adult worm (male and female if available), ova, ciliated miracidium, the sporocyst in the snail tissue, and the infective ciliate cercaria.

    b. Next read this brief summary of the natural history of *Schistosoma mansoni* (figure 20.5 and **Nester et al. Microbiology: A human perspective, 7th ed. The McGraw-Hill Companies, 2012, p. 350).**

Human schistosomiasis occurs wherever these conditions exist: water is polluted with human wastes; this water is used for human bathing and wading, or irrigation of cropland; and the presence of snail species that are necessary as hosts for the sporocyst stage in fluke development and completion of its life cycle. The solution to this public health problem is very complex, not only because of technical difficulties in its control and treatment, but also because its life cycle presents an ecological dilemma. Many developing countries need food desperately, but the main sources now available for these expanding needs are fertile deserts, which have adequate nutrients but require vast irrigation schemes, such as the Aswan Dam in Egypt. However, due to the unsanitary conditions and the presence of suitable snail hosts, these projects are accompanied by an increase in the disease schistosomiasis, which currently is very difficult to control and very expensive to treat on a wide scale.

The **cercaria** larvae swim in the contaminated water, penetrating the skin of barefoot agricultural workers. They migrate into the blood and collect in the veins leading to the liver. The *adults* develop there, mate, and release the *eggs*. The eggs are finally deposited in the small veins of the large intestine, where their spines cause damage to host blood vessels. Some eggs die; however, others escape the blood vessels into the intestine and pass with the feces into soil and water. There they develop, and then hatch into motile **miracidia,** which eventually infect suitable snail hosts and develop into saclike **sporocysts** in the snail tissues. From this stage, the fork-tailed cercaria larvae develop; these leave the snail and swim in the water until they die or find a suitable human host, thus completing the complex life cycle involving two hosts and five separate stages.

c. Now look at the prepared slides of all the *Schistosoma* stages discussed in the preceding description. Sketch each stage in the appropriate place in the life cycle diagram shown in part 5 of the Laboratory Report.

EXERCISE

20

Laboratory Report: Parasitology: Protozoa and Helminths

Results

1. Examination of Free-Living Cultures
   a. Pond water examination.
      Description of movements and drawings of any protozoans found in pond water.

   b. Examination of fresh samples of a free-living amoeba, paramecium, and flatworm.
      Description of movements and drawings with labels.

2. Examination of Stained Slides for Trophozoites and Cysts

Prepare drawings of the trophozoite and cyst stage of either *Entamoeba histolytica*, *Giardia lamblia*, or *Balantidium coli*. Label accordingly (see Procedure step 2).

3. Examination of Protozoans Present in Stained Blood Slides

a. Examine blood smears of *Plasmodium vivax* (see Procedure step 3*a*).

b. Blood smear of *Trypanosoma gambiense* (see Procedure step 3*b*).

4. Comparison of a Free-Living Worm with Its Parasitic Relative

a. Comparison of *Dugesia* or *Planaria* species (free-living) with *Clonorchis sinensis* or *Fasciola hepatica* (parasitic). See Procedure step 4*a*.

b. Study of a parasitic tapeworm (*Taenia* species). See Procedure step 4c.

c. Study of *Ascaris lumbricoides*. See Procedure step 4d.

d. Study of *Dirofilaria immitis*. See Procedure step 4e.

5. Life Cycle of *Schistosoma mansoni* and Possible Methods of Control
   a. For each space in this life cycle, sketch the appropriate stage, using the prepared microscope slides.

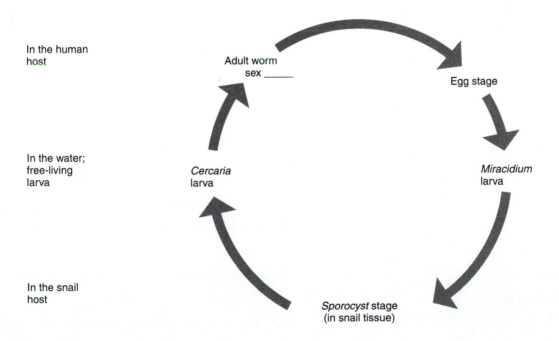

b. Propose a plan for public health control of schistosomiasis. Describe various strategies that might be developed by public health personnel to interrupt this cycle and thus prevent schistosomiasis. Show on a diagram where specific measures might be taken, and label. Explain each possibility and its advantages and disadvantages.

Questions

1. Which form—the trophozoite or the cyst—is most infective when found in a feces sample? Explain.

2. In what ways are free-living and parasitic worms similar, such that they can be identified as closely related?

3. In what ways do the parasitic species differ from the free-living planaria? Use the chart to summarize your comparisons.

	Planaria	Fluke	Tapeworm
Outside covering			
Organs of attachment			
Sensory organs			
Digestive system			
Reproduction			

4. Estimate the length and width of a trypanosome. See color plate 17 for a clue. Show your calculations.

5. How is the *Echinococcus* tapeworm transmitted to humans? Does it cause a serious disease? What are two ways in which its transmission to humans can be prevented?

INTRODUCTION to Medical Microbiology

It is easy to think of microorganisms as deadly, vicious force—especially when the diseases they cause kill young people or wipe out whole populations. The organisms, however, are simply growing in an environment they find favorable.

If a pathogen becomes too efficient at taking advantage of its host, the host dies and the organism dies with it. Thus, the most successful pathogens are those that live in balance with their hosts. When a new pathogen enters the population, it is very virulent, but after awhile there is a selection toward less virulent pathogens and also a selection in the hosts for increased resistance.

Medical microbiology continues to offer challenges to those interested in medicine and in pathogenic bacteria. These next exercises are an introduction to many of the organisms that are encountered in a clinical laboratory. Not only will you study the characteristics of the organisms, but also you will learn some strategies for isolating and identifying them. In addition, these exercises are designed to help you learn to differentiate between organisms you can expect to find as normal flora in various places in the body and others that are responsible for certain diseases.

22

Normal Skin Biota

Getting Started

The organisms growing on the surfaces and in the orifices of the body are called *normal biota*. They are usually considered **commensals** because they do not harm their hosts; in fact, they have several beneficial roles. Normal biota prevent harmful organisms from colonizing the skin because they are already established there and utilize the available nutrients. Some produce enzymes or other substances that inhibit nonresident organisms. Other organisms, called transients, can also be found on the skin for short periods, but they cannot grow there and soon disappear.

Familiarity with organisms making up the skin biota is useful because these organisms are frequently seen as contaminants. Skin is continually flaking off, and bacteria floating in the air on rafts of skin cells sometimes settle into open petri dishes. If you are familiar with the appearance of *Staphylococcus* and *Micrococcus* colonies, for instance, you will suspect contamination if you see such colonies on an agar plate.

One way of distinguishing some skin biota from each other is by using an anaerobe container (figure 22.1) as some bacteria can grow, whereas others cannot. Different packets can be purchased to place inside the container that create carbon dioxide (CO_2) and hydrogen ions (H^+). Palladium pellets placed in the container catalyze the H^+ binding to the oxygen (O_2) present to form

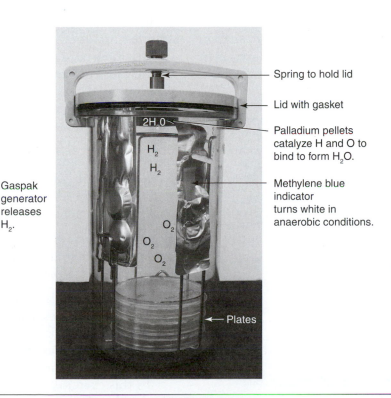

Spring to hold lid

Lid with gasket

Palladium pellets catalyze H and O to bind to form H_2O.

Methylene blue indicator turns white in anaerobic conditions.

$2H_2O$

H_2

H_2

O_2

O_2

O_2

Gaspak generator releases H_2.

Plates

Figure 22.1 An anaerobe jar showing the components required to obtain an anaerobic state. Other systems can be used instead that utilize a sealed plastic bag and tablets to create and show the anaerobic state.

Courtesy of Anna Oller, University of Central Missouri.

nontoxic water (H_2O). A methylene blue or other indicator can be added to the container before it is sealed to visually ensure than an anaerobic state has been created inside of the container. A methylene blue strip appears blue in the presence of oxygen but turns clear in the absence of oxygen. The container must be sealed after adding the petri dishes, indicator, and packets to prevent oxygen from re-entering the container.

Some of the organisms you may isolate include the following (figure 22.2):

Staphylococcus epidermidis This Gram-positive coccus is a facultative anaerobe found on the skin as part of the normal biota of almost all humans. It is also isolated from many animals and is coagulase and mannitol salt negative.

Staphylococcus aureus This Gram-positive coccus is a facultative anaerobe found on the skin and nares (nostrils), appearing white to light yellow, depending on the growth medium. At least 20% of the population possess this bacterium, which is identified by a positive **coagulase test** and mannitol salt plate. It seems to cause no harm to its host but is frequently the cause of wound infections and food poisoning.

Methicillin-resistant *Staphylococcus aureus* (MRSA) has gained attention recently. It is resistant to the antibiotics methicillin and oxacillin, and it

tests positive for the **beta-lactamase** test. Wounds infected by MRSA are extremely painful and can swell to the size of a golfball overnight. Recently, many additional species of *Staphylococcus*, associated mostly with diseases in immunologically compromised individuals, have been identified.

Tests used to distinguish *Staphylococcus aureus* from *S. epidermidis* include a blood agar plate (exercise 23), a coagulase test, and a mannitol salt plate. *Staphylococcus aureus* is β-hemolytic on a blood agar plate, whereas *S. epidermidis* is either γ- or α-hemolytic, depending on the strain. The coagulase test determines if the bacterium produces the enzyme coagulase, which clumps plasma. The mannitol salt plate uses 7.5% salt to inhibit most Gram-negative bacteria, and it contains a phenol red pH indicator that turns the red media yellow if the sugar mannitol is fermented (color plate 25).

Micrococcus luteus This Gram-positive coccus is an aerobe found on the skin of some people, but it almost never causes disease. It is frequently an air contaminant forming bright yellow colonies.

Propionibacterium acnes These anaerobic, Gram-positive rods are **diphtheroid,** or club shaped, and colonies are usually a few mm in diameter. When investigators tried to isolate an organism that might be the cause of **acne,** they almost always found the same Gram-positive diphtheroid rods in the lesions. Therefore, they named the organism *Propionibacterium acnes*. However, when people without acne were studied, it was found that *P. acnes* was present on their foreheads as well. Although some people have a much higher population of this organism than others, the elevated number does not seem to correlate with acne or any other skin condition.

Propionibacterium granulosum A Gram-positive, anaerobic diphtheroid rod found in smaller numbers than *P. acnes*. It is considered a harmless commensal.

Bacillus subtilis and *Bacillus cereus* *Bacillus* species are facultative anaerobes, can be found on skin as transients or commensals, and normally do not cause disease. *Bacillus* colonies often have a rough appearance (color plate 1 similar to the texture of colony 6).

In this exercise, you will grow your normal skin biota and perform various tests to identify the colonies that grew. You may also investigate if *P. acnes* correlates to individuals reporting acne.

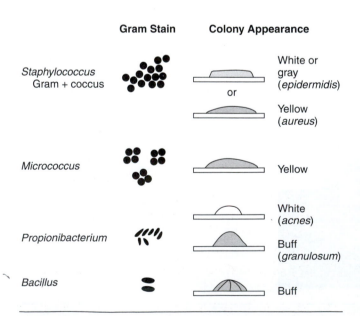

Figure 22.2 Colonial appearance of some normal skin biota organisms in a Gram stain.

Definitions

Acne. A skin condition characterized by white-heads, blackheads, and infected hair follicles.

Beta-lactamase. An enzyme that degrades the beta-lactam ring of antibiotics like penicillin and methicillin. Bacteria that produce this enzyme will test positive for beta-lactamase.

Coagulase test. A test in which organisms are mixed with plasma on a slide. If the cells clump together, the culture is coagulase positive.

Commensals. Organisms that live together in close association and may or may not benefit each other.

Diphtheroid. A Gram-positive, club-shaped organism sometimes called a *coryneform*. *Propionibacterium* and *Corynebacterium* are examples of diphtheroid-shaped organisms.

Objectives

1. Differentiate organisms making up the normal skin biota.
2. Describe the importance of skin biota.
3. Describe how the anaerobe jar works.

Prelab Questions

1. What is the purpose of rubbing your forehead with ethanol?
2. Why would you incubate plates both aerobically and anaerobically?
3. Which *Staphylococcus* is coagulase positive and why is this test important?

Materials

First Session
   TSY (trypticase soy yeast) agar plates (or TSY contact plates, 2" in diameter), 2
   GasPak anaerobe jar (or other anaerobic system)
Per student
   Cotton balls
   70% ethanol
   Sterile swab, 1
   Sterile saline
   Mannitol Salt Plate, 1

Optional
   Sterile swab, 1
   TSY agar plates for dilution, 2
   5.0 ml sterile saline, 1
   9.9 ml sterile saline blank, 1
   1 ml sterile pipets, 3
   Sterile bent glass rod (sometimes called dally rods or hockey sticks), 1 or 2
Second Session
   TSY + glucose agar deeps (yeast extract and glucose are added to trypticase soy agar to encourage *Propionibacterium* growth), 4
   TSY + glucose + bromcresol purple agar slants, 2
   Gram-stain reagents
   Plasma
   H_2O_2
   Magnifying glass is optional but helpful
Third Session
   Plasma (optional)

PROCEDURE

Safety Precautions: Some students may isolate *Staphylococcus aureus* as part of their normal biota. This is a pathogen and should be handled with extra care.

First Session

1. Saturate a cotton ball with 70% ethanol and rub your forehead for 20 seconds. This will remove any transient skin organisms (those that are not part of your normal biota).
2. Let your forehead dry for about 15 minutes. Avoid touching it with your hair or fingers.
3. Moisten a sterile swab with saline and rub it briskly on an area of your forehead about the size of a quarter for about 15 seconds.

4. Immediately roll the swab over the first third of one TSY agar plate; then using the same swab, roll the swab over the first third of a second TSY agar plate, and thirdly onto the first third of a mannitol salt plate. Discard the swab, and finish the streak plates with a loop.

Alternative method to steps 3 and 4:
Press an open contact TSY agar plate to your forehead. Repeat with subsequent plates on an adjacent area.

5. Incubate one TSY agar plate and the mannitol salt plate aerobically at 37°C for 24–48 hours.

6. Be sure to label the plates as aerobic and anaerobic incubation.

7. Incubate the second TSY agar plate anaerobically in an anaerobe container at 37°C for 4–5 days. Follow the manufacturer's directions for creating an anaerobic atmosphere.

8. After 24–48 hours of incubation, you or the instructor should store the aerobic plates in the refrigerator to prevent the plates from drying out. *Staphylococcus* and *Micrococcus* can be observed after 24 hours, but *Propionibacterium* must be incubated 5 days before colonies can be seen.

Optional

1. Moisten a 2nd sterile swab with saline and rub it briskly on an area of your forehead about the size of a quarter for about 15 seconds.

2. Rinse the swab in a 5 ml tube of saline. Discard the swab as directed. Mix the sample thoroughly.

3. Pipet 0.1 ml from the 5 ml saline tube into a 9.9 ml saline blank and mix thoroughly. Discard the pipet as instructed.

4. Remove 0.1 ml from the 9.9 ml saline tube and place it on the surface of a TSY agar plate. Label the plate as TSY diluted saline.

5. Pipet 0.1 ml from the 5 ml saline tube and plate it on the surface of a 2nd TSY agar plate. Label the plate as TSY from undiluted saline. Discard the pipet as instructed.

6. With a sterile bent glass rod, spread the 0.1 ml solution over the agar surface containing the

diluted sample and then over the surface of the TSY with the undiluted sample. Be careful to avoid contaminating the rod between plates.

7. Place the rod in a container to be washed and sterilized as directed by the instructor.

8. Incubate the plates anaerobically at 37°C for 4–5 days.

Second Session (5 Days Later)

Note: For each TSY colony you examine, you will perform several tests (Gram stain, glucose slant, shake tube, catalase test) so using a needle is recommended.

1. Examine the aerobic TSY plate and circle two different colony types with a marking pen on the bottom of the plate. Make a Gram stain of part of each circled colony.

2. If the colonies are Gram-positive cocci, inoculate
   a. a glucose + bromcresol purple TS agar slant
   b. a cooled melted agar deep/shake tube (exercise 9). Incubate at 37°C for 24–48 hours.

3. Examine the anaerobic TSY plate. You will see some of the same colony types observed on the aerobic plate. (Your instructor may have you use the optional diluted and undiluted plates.)

4. Choose two possible *Propionibacterium* colonies and Gram stain them.

5. If they are diphtheroid Gram-positive rods, inoculate into a shake tube and perform a catalase test (figure 23.3a). Incubate the shake tube at 37°C for 5 days.

6. Examine the mannitol salt plate. Count the number of colonies that tested positive and negative for mannitol fermentation and record in the Laboratory Report.
   a. Circle two different colony types with a marking pen on the bottom of the plate and label them 1 and 2. You should not need to make a Gram stain, as organisms that grew on this agar should be Gram positive.

b. Perform a coagulase test on each colony you circled. Do you have any *Staphylococcus aureus* present?

Optional

7. From the TSY plate inoculated from the saline tubes, estimate the number of *Propionibacterium acnes* on the plates. (Assume all small white colonies are *P. acnes*.) If the numbers are very large, divide the plate in fourths and count one section. Record as TNTC (too numerous to count) if you cannot distinguish separate colonies. If you do not see any, record as TFTC (too few to count). Calculate the estimated number of *Propionibacterium acnes* from the swabbed area:

$$1/10\text{ml} \times 10 \times \#\text{ colonies} = \text{cfu/ml}$$

The student can anonymously give the results of his or her forehead sample and indicate whether he or she has acne. The instructor can compile the class results and discuss how the presence of *P. acnes* and the condition correlate in this very small sample.

8. On a separate piece of paper, provide the estimated number of *P. acnes* from the area sampled and indicate if you now have acne or are being treated for acne (yes or no). Hand the paper into the instructor so that the data can be posted. Please do not sign your name to the paper.

9. Record the results.

10. Discuss if the class results support the original observation: *P. acnes* is a component of almost everyone's skin biota, whether acne is present or not.

11. It is important to note that these results can only be described as preliminary because the sample size is very small. Also, for a study to have validity, the techniques must be carefully standardized.

Third Session (5 Days Later)

1. Observe the glucose + bromcresol purple slants. If the organism is able to ferment glucose, the acid produced will turn the purple agar yellow.

Coagulase Test

Method 1

1. Place a drop of water on a slide and make a *very* thick suspension of cells from a yellowish colony (not bright yellow).
2. Place a drop of plasma next to it, and mix the two drops together. Look for clumping; clumped cells indicate a coagulase-positive result.
3. Drop the slide in boiling water, and boil for a few minutes to kill the organisms before cleaning the slide.

Method 2

1. Mix a loopful of yellowish cells with 0.5 ml undiluted rabbit plasma in a small tube. The plasma can also be diluted 1:4 with saline.
2. Label the tube. Incubate at 37°C, and examine after 4–24 hours. Tip the tube sideways. A solid clot is a positive test (color plate 26).

2. Observe the agar deeps. Obligate aerobes are only able to grow on the top, whereas facultative anaerobes can grow throughout the entire tube. The obligate anaerobes cannot grow in the top few centimeters where oxygen has diffused—only in the bottom anaerobic portion (see exercise 9).

3. Record your results and identify your isolates. The table lists a description of the organisms most commonly found on the forehead.

Optional

4. If you have yellow colonies of *Staphylococcus*, you can determine if the species is *S. aureus* with a coagulase test. *S. aureus* is coagulase positive, and *S. epidermidis* is coagulase negative.

	Summary of Reactions			
	Gram Stain	**Colony Color on TSY**	**Metabolism**	**Glucose**
Staphylococcus epidermidis	+ cocci, clusters	white/gray	facultative	+
Staphylococcus aureus	+ cocci, clusters	yellow	facultative	+
Micrococcus	+ cocci, tetrads	bright yellow	obligate aerobe	−
Propionibacterium	+ rods, palisades	small white, pink/buff	obligate anaerobe/ aerotolerant	+
Bacillus	+ rods (large)	white/buff	facultative	+

EXERCISE

22

Laboratory Report: Normal Skin Biota

Results

	Isolate 1	Isolate 2	Isolate 3	Isolate 4
TSY Plate Colony Isolate Results				
1. Colony appearance				
2. Gram stain				
3. Glucose fermentation				
Agar Deep				
4. Catalase				
5. Possible identity				

Although this was not a quantitative procedure, what organism seemed to be the most numerous on your forehead?

Mannitol Salt Plate

	Numbers Present			Numbers Present
Mannitol positive		Mannitol negative		
	Colony Isolate 1			**Colony Isolate 2**
Colony appearance				
Mannitol reaction				
Coagulase				
Possible identity				

1. How many *P. acnes* did you isolate from the swabbed area of your forehead? _____

Optional Class Results

2. Number of students in the class. _____
3. Number of students isolating *P. acnes* from his or her forehead. _____
4. Number of students with self-described acne. _____
5. Number of students with large numbers of *P. acnes* and acne. _____
6. Number of students with large numbers of *P. acnes* and no acne. _____
7. Conclusion:

Questions

1. How could normal skin biota be helpful to the host?

2. How can you immediately distinguish *Staphylococcus* from *Propionibacterium* in a Gram stain?

3. Why does *Staphylococcus* probably cause more contamination than *Propionibacterium*, even though most people have higher numbers of the latter?

    Hint: Are most agar plates incubated aerobically or anaerobically?

4. Prepare a flow chart that would help you identify your skin biota isolates.

EXERCISE 25

Clinical Unknown Identification

Getting Started

In this exercise, you will have an opportunity to utilize the knowledge and techniques you have learned in the previous three exercises to identify unknown organisms. You are given a simulated (imitation) **clinical specimen containing two organisms,** and your goal is to separate them into two pure cultures and identify them using different media and tests. The organisms are either associated with disease or are common contaminants found in normal biota or the environment.

Your unknown specimen represents a urine infection, a wound infection, or food poisoning. In actual clinical cases, standardized procedures exist for each kind of specimen. However, you will be identifying only a limited number of organisms. With some careful thought, you can plan logical steps to use in identifying your organisms.

The following are characteristics useful in identifying your unknown organism.

Bacterial Cell Morphology The size, shape, arrangement, and Gram-staining characteristics of the bacteria as determined by the Gram stain. It also may include the presence of special structures, such as endospores.

Colonial Morphology The appearance of isolated colonies on nutrient (complete) media, such as TS agar or blood agar, including their size, shape, and consistency.

*Growth on **Selective Media*** The ability of organisms to grow on selective media. Mannitol salt, which selects for organisms tolerating 7.5% salt; eosin methylene blue (EMB), which selects for Gram-negative organisms; and MacConkey, which selects for Gram-negative rods, are examples of selective media.

*Reactions on **Differential Media*** The color of colonies on EMB agar or MacConkey agar is based on lactose fermentation (lactose fermenters are maroon or dark pink, respectively). The appearance of organisms on mannitol salt agar is based on mannitol fermentation (mannitol fermenters turn the medium yellow). The presence of hemolysis on blood agar constitutes another type of reaction of bacterial enzymes on red blood cells.

Biochemical Capabilities These include the ability to ferment different carbohydrates and the production of various end products, as well as the formation of indole from tryptophan. Tests include methyl red, Voges–Proskauer, citrate utilization, urease, catalase, oxidase, and coagulase.

Approach the identification of your "unknown" clinical specimen with the following steps. Your instructor will tell you which approach (you were given pure cultures or you were given two unknowns that you have to isolate) you will use:

1. Make a Gram stain of the specimen.

2. Streak the cultures on the appropriate complete and selective medium.

3. After incubation, identify two different colony types and correlate with their Gram reaction and shape. Also correlate the growth and appearance of the colonies on selective media with each of the two organisms.

4. Restreak for isolation. It is useless to do any identification tests until you have pure cultures of the organisms.

5. After incubation, choose a well-isolated colony and inoculate a TS agar slant to be used as your stock culture. Prepare a stock culture for each organism.

6. Inoculate or perform various tests that seem appropriate. Keep careful records. Record your results on the worksheets as you observe them.

7. Identify your organisms from the test results.

Definitions

Clinical specimen. Cultures encountered in a medical laboratory.

Differential media. Media that permit the identification of organisms based on the appearance of their colonies.

Selective media. Media that permit only certain organisms to grow and that aid in isolating one type of organism in a mixture of organisms.

Note: See exercises 22–24 for more information on these organisms and tests.

Objectives

1. Apply your knowledge to a microbiological problem.
2. List the procedures used to isolate and identify clinical specimens.
3. Differentiate the presence of contaminants or nonpathogens in clinical specimens from the bacteria that should be present.

Prelab Questions

1. Why is it important to do a Gram stain first on your unknown?
2. Why is it important to perform an isolation streak on your unknown?
3. Name a bacterium from this lab that will be catalase negative.

Materials

First Session

24-hour unknown cultures (mixture or pure culture(s)) labeled with hypothetical source (for each student or team of two students)—plates, slants, or broth tubes

For a **mixed** culture:

Blood agar plate or trypticase soy agar plate, 1 per student

MacConkey agar plate (or EMB agar plate)

Mannitol salt agar plate

Gram-staining reagents

For a **pure** culture:

All tubes and plates listed; and oxidase, H_2O_2, and staining reagents

Second and Third Sessions

Trypticase soy agar plates

Bile esculin agar slants

Trypticase Soy broths with 6.5% NaCl

Nutrient agar slants

Citrate agar slants

Urea slants

Glucose + bromcresol purple agar slants

Fermentation broths of glucose, lactose, and sucrose (or TSI slants)

MR-VP broths for the methyl red and Voges–Proskauer test

Tryptone broths for the indole test (or SIM tubes)

Shake agar deeps or thioglycollate broth tubes (oxygen requirements)

XLD plates

PAD plates or slants

HE (or SS) plates (optional)

Reagents

Oxidase reagent

Kovacs reagent

Voges–Proskauer reagents A and B

Methyl red for methyl red test

Plasma for coagulase test

H_2O_2

12% ferric chloride

Optional: Staining material for the endospore, acid-fast, and capsule stains

PROCEDURE

First Session

Your instructor will tell you if you were given a mixed or pure culture.

If you are given a **mixed** culture:

1. Make a Gram stain of the culture. Observe it carefully to see if you can see both organisms. Record your result in the Laboratory Report. You can save the slide and observe it again later if you have any doubts about it. You can also save the broth, but one organism may overgrow the other.

2. Inoculate a complete medium agar plate such as trypticase soy agar or blood agar, and appropriate selective and differential agar plates. Use MacConkey agar (if you suspect the possibility of a Gram-negative rod in a urine specimen) or a mannitol salt agar plate (if you suspect *Staphylococcus* in a wound infection). Streak the plates for isolated colonies.

3. Incubate at 37°C for 24–48 hours.

If you are given a **pure** culture:

1. Make a Gram stain of the culture. Record your result in the Laboratory Report. You can save the slide and observe it again later if you have any doubts about it.

2. Inoculate a complete medium agar plate such as trypticase soy agar or blood agar, oxygen shake tubes, and appropriate selective and differential tests (based on your Gram stain, cell shape, and cell arrangement). Streak plates for isolated colonies.

3. Incubate at 37°C for 24–48 hours.

Second Session

If you were given a **mixed** culture:

1. Examine the streak plates after incubation and identify the two different colony types of your unknown organisms, either on the complete medium or the selective media, wherever you have well-isolated colonies. Gram stain each colony type (organisms usually stain better on nonselective media). Also identify each colony type on the selective and differential media so that you know which organisms grow on the various media. Record their appearance on the differential media as well. It is helpful to circle colonies that you Gram stain on the bottom of the petri plate with a marking pen.

2. Inoculate and perform other appropriate tests on your isolated microbes.

3. Restreak each organism on a complete medium (instead of selective media) for isolation. This technique ensures that all organisms will grow and that you will be able to see if you have a mixed culture. Do not discard the original streak plates of your isolates—store at room temperature. If at some point your isolate does

not grow, you will be able to go back to the old plates and repeat the test.

If you were given a **pure** culture:

1. Examine your tubes and plates and record your test results in the Laboratory Report.

2. Finish any stains or tests that you need to perform on your unknown.

Third Session

If you were given a **mixed** culture:

1. Observe the plates after incubation. If your organisms seem well isolated, inoculate each one on a TS agar slant to use as your stock culture. If you do not have well-isolated colonies, re-streak them. It is essential that you have a pure culture. Possible steps in identifying Gram-positive cocci follow.

2. Look at the possible list of organisms and decide which ones you might have, based on the information you have found so far.

These are a few of the test results. Your instructor may provide more. Plan work carefully and do not waste media using tests that are not helpful. For example, a urea slant would not be useful for distinguishing between *Staphylococcus epidermidis* and *Staphylococcus aureus*.

Simulated wounds
  Staphylococcus epidermidis
  Staphylococcus aureus
  Micrococcus luteus
  Pseudomonas aeruginosa
  Mycobacterium phlei
Simulated urine infection
  Escherichia coli
  Enterobacter aerogenes
  Proteus
  Enterococcus faecalis
  Streptococcus bovis
  Klebsiella pneumonia
  Alcaligenes faecalis
  Citrobacter freundii
  (plus wound organisms)
Simulated food poisoning
  Bacillus cereus
  Salmonella enteriditis
  Shigella flexneri

A partial listing of the unknown organisms' possible identities follows. Others can be added.

Gram-Positive Cocci (Nonmotile)	
Staphylococcus aureus Found in urine, wounds, or food	Gram-positive cocci in grapelike clusters (bunches), yellowish colony, catalase positive, ferments glucose (acid) and mannitol, coagulase positive, salt tolerant
Staphylococcus epidermidis Contaminant	Gram-positive cocci in grapelike clusters, white colony, catalase positive, ferments glucose (acid) but not mannitol, coagulase negative, salt tolerant
Micrococcus Contaminant	Gram-positive cocci in tetrads, yellow colony, catalase positive, does not ferment glucose or mannitol, coagulase negative, salt tolerant, aerobe, bile esculin negative
Enterococcus faecalis Found in urine	Gram-positive cocci in chains, catalase negative, ferments glucose (acid), coagulase negative, bile esculin positive, salt tolerant (turbid)
Streptococcus bovis Contaminant	Gram-positive cocci in chains, catalase negative, ferments glucose (acid), coagulase negative, bile esculin positive, not salt tolerant

Gram-Positive Rod	
Bacillus Found in food or contaminant	Large Gram-positive rods, forms endospores, catalase positive, ferments mannitol

Gram-Negative Rods	
Escherichia coli Found in urine or food	Glucose & lactose positive (acid and gas), indole positive, methyl red positive, Voges–Proskauer negative, citrate negative, urea negative, oxidase negative, sulfur negative, PAD negative, MAC, XLD, HE, SS positive, motile, TSI: A/A g
Proteus vulgaris Found in urine	Glucose & sucrose positive (acid), lactose negative, indole positive, methyl red positive, Voges–Proskauer negative, citrate negative, urea positive, oxidase negative, sulfur positive, PAD positive, MAC, XLD, HE, & SS positive, motile, TSI: A/A+
Pseudomonas Found in urine and wounds	Glucose & lactose negative, indole negative, citrate positive, urea negative, oxidase positive, sulfur negative, motile, TSI: K/K, aerobe, blue-green or yellow pigments, smells like grapes
Enterobacter Found in urine	Glucose, sucrose, & lactose positive (acid and gas), mannitol positive, indole negative, methyl red negative, Voges–Proskauer positive, citrate positive, urea negative, oxidase negative, MAC, XLD, HE, & SS positive, motile, TSI: A/A g, capsules
Citrobacter Found in urine	Glucose and lactose positive (acid), mannitol positive, indole negative, methyl red positive, Voges–Proskauer negative, citrate positive, urea negative, sulfur positive, PAD negative, XLD positive, MAC, XLD, HE, & SS positive, motile, TSI: A/A+
Alcaligenes faecalis Found in urine and wounds	Glucose and lactose negative, indole negative, citrate negative, urea negative, oxidase positive, motile, TSI: K/K, non-pigmented
Klebsiella Found in urine	Glucose & lactose positive (acid and gas), indole negative, methyl red positive or negative, Voges–Proskauer negative, citrate positive, urea negative, oxidase negative, sulfur negative, PAD negative, XLD, HE, & SS positive, nonmotile, mucoid colonies, TSI: A/A g, capsules
Salmonella Found in food	Glucose positive (acid), indole negative, methyl red positive, Voges–Proskauer negative, citrate positive, urea negative, sulfur positive, PAD negative, MAC, XLD, HE, and SS negative, motile, TSI: K/A+
Shigella Found in food	Glucose positive (acid), indole positive, methyl red positive, Voges–Proskauer negative, citrate negative, urea negative, sulfur negative, PAD negative, MAC, XLD, HE, and SS negative, nonmotile, TSI: K/A

Other	
Mycobacterium	Acid-fast rods, rough colonies, aerobe, citrate positive, nonmotile

EXERCISE

25

Laboratory Report: Worksheet
and Final Report: Clinical
Unknown Identification

Gram stain of original specimen: _____

_____

(Describe cell shape, arrangement, and Gram reaction)

Gram stains of TS agar subcultures: _____

_____

(Describe cell shape, arrangement, and Gram reaction)

Test	Organism #1	Organism #2
Colony description		
(TS agar or blood)		
Hemolysis		
Gram stain		
Colony appearance MacConkey (or EMB)		
Colony appearance mannitol salt agar		
Special stains		
capsule		
endospore		
acid fast		
Lactose fermentation		
Glucose fermentation		
Sucrose fermentation		

Test	Organism #1	Organism #2
TSI slant		
Mannitol fermentation		
Sulfur production		
Indole production		
Motility		
Methyl red		
Voges–Proskauer		
Citrate utilization		
Urea hydrolysis		
Catalase test		
Coagulase		
Oxidase		
Oxygen requirements		
XLD		
Hektoen Enteric (or SS)		
PAD		

Consult the table compiled from the class results at the end of the second session in exercise 24 and figure 23.1 to help you identify your organisms.

Final Identification
Organism #1 = _____

Organism #2 = _____

Final Report

1. What is the identification of your organisms? Discuss the process of identification (reasons for choosing specific tests, any problems, and other comments).

   Organism #1:

   Organism #2:

INTRODUCTION to Some Immunological Principles and Techniques

Immunology, the study of the body's immune response, is responsible for protecting the body against disease. It is often triggered when foreign substances or organisms invade the body. Examples include pathogenic microbes and the chemical compounds they produce, such as foreign materials called **antigens.** In diseases such as AIDS and cancer, the body's immune response is either seriously weakened or destroyed, whereas for milder diseases the immune response remains complete.

Different forms of the immune response include **phagocytic cells,** such as white blood cells (WBCs); **enzymes,** such as **lysozyme;** and **antibodies.** WBCs and enzymes are examples of natural immunity because they are already present in the body and need not be triggered by the antigen. In contrast, antibodies are an example of acquired immunity because their formation is triggered only in the presence of the antigen. For some people, the immune response can be triggered by the body's own proteins. This can result in the formation of **autoimmune diseases,** such as rheumatoid arthritis and glomerulonephritis.

Phagocytic cells and enzymes are also examples of nonspecific immunity because they can react with a variety of different foreign substances (for instance, phagocytic cells can engulf both inanimate and animate particles). Conversely, antibodies represent a form of specific immunity because they are produced in response to particular antigens (an antibody produced against *Salmonella* cell walls will not react with *Proteus* cell walls).

The exercises include examples of both innate immunity (exercises 26 and 27) and acquired immunity (exercise 28). In exercise 26, you will study human blood cells and learn how to determine which ones are phagocytic. In exercise 27, you will learn how to determine the bactericidal activity of the enzyme lysozyme, which occurs naturally in phagocytic white blood cells, saliva, nasal secretions, and tears. Lysozyme is able to digest the cell walls of many bacteria.

Exercise 28 represents the current techniques for demonstrating particulate antigen–antibody reactions using the enzyme-linked immunosorbent assay (ELISA), where you will use it for identification of a sexually transmitted disease. It is also widely employed in plant and animal virus identification.

Definitions

Antibody. A protein produced by the body in response to a foreign substance (e.g., an antigen); it reacts specifically with that substance.

Antigen. Any cell particle or chemical that can cause production of specific antibodies and combine with those antibodies.

Autoimmune disease. An immune reaction against one's own tissues.

Enzyme. A protein that acts as a catalyst. A catalyst is a substance that speeds up the rate of a chemical reaction without being altered or depleted in the process.

Lysozyme. An enzyme that degrades the peptidoglycan layer of the bacterial cell wall.

Phagocytic cells. Cells that protect the host by ingesting and destroying foreign particles, such as microorganisms and viruses.

Some of our cells, although they are part and parcel of us, have not even fixed coherence within our "rest." Such cells are called "free" . . . . The cells of our blood are as free as fish in a stream. Some of them resemble in structure and ways so closely the little free swimming amoeba of the pond as to be called amoeboid. The pond amoeba crawls about, catches and digests particles picked up in the pond. So the amoeboid cells inhabiting my blood and lymph crawl about over and through the membranes limiting the fluid channels in the body. They catch and digest particles. Should I get a wound, they contribute to its healing. They give it a chance to mend, by eating and digesting bacteria which poison it and by feeding on the dead cells which the wound injury has killed. They are themselves unit lives and yet in respect to my life as a whole, they are components in that corporate life.

Sherrington, *Man on His Nature*

26

Differential White Blood Cell Stains

Getting Started

This exercise examines the cellular forms of the immune system, specifically the white blood cells (WBCs). For the most part, they can be distinguished from one another using a blood smear stained with Wright's stain, a differential stain. This stain uses a combination of an acid stain, such as eosin, and a basic stain, such as methylene blue. They are contained in an alcoholic solvent (methyl alcohol), which fixes the stains to the cell constituents, particularly since the basophilic granules are known to be water soluble. With this stain, a blood smear shows a range in color from the bright red of acid material to the deep blue of basic cell material. In between are neutral materials that exhibit a lilac color. There are also other color combinations, depending upon the pH of the various cell constituents.

The two main groups of WBCs are the granulocytes (cytoplasm contains granules) and the agranulocytes (clear cytoplasm). The WBCs, or leukocytes, represent approximately 1/800 of the total blood cells. Common cell types found in the granulocytes are **neutrophils, eosinophils,** and **basophils.** The granulocytes are highly phagocytic and contain a complex, segmented nucleus. The granules in neutrophils stain light pink, red in eosinophils, and purple in basophils. Eosinophils often have a nucleus with two lobes, whereas the other granulocytes have four or more lobes. The agranulocytes are relatively inactive and have a simple nucleus, or kidney-shaped nucleus. The basic agranulocyte cell types are the **lymphocytes** and **monocytes.** Lymphocytes are smaller cells with little cytoplasm, and monocytes are large cells with more cytoplasm that often contains cytoplasmic vacuoles. Another WBC type found is the **platelet** (very small, multinucleate, irregular pinched off parts of a **megakaryocyte**). Platelets are responsible for clotting the blood to prevent excessive bleeding. The appearance of these cell types in blood stained with a differential stain is illustrated in figure 26.1. Many of the leukocytes are **amoeboid,** capable of moving independently through the bloodstream. They also move out into the tissues, where they repel infection and remove damaged body cells from bruised or torn tissue. They are defensive cells, specialized in defending the body against infection by microorganisms and other foreign invaders.

Differential blood stains are important in disease diagnosis because certain WBCs either increase or decrease in number depending on the disease. In making such judgments, it is important to know the appearance of normal blood (color plate 48). The microscopic field shown in the color plate includes mostly RBCs with a neutrophil, an eosinophil, a

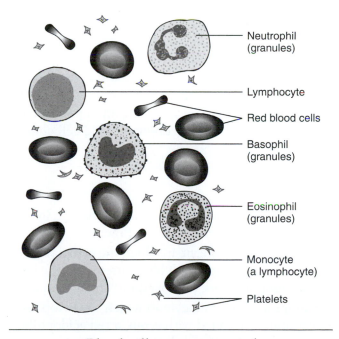

Figure 26.1 Blood cell types present in human peripheral blood. The granular leukocyte names find their origin from the color reaction produced by the granules after staining with acidic and basic components of the staining solution. Neutrophil = neutral-colored granules; basophil = basic color; and eosinophil = acid color.

lymphocyte, and some platelets. Neutrophils are often increased in bacterial infections, eosinophils are often increased in parasitic infections (and allergies), and lymphocytes are often increased in viral infections. A Giemsa stain can be used to see *Plasmodium* species in RBCs, but they cannot be seen with only the Wright stain. In a differential performed on human blood, you should see approximately 40–75 neutrophils, 13–43 lymphocytes, 2–12 monocytes, 1–4 eosinophils, and 1 basophil.

The red blood cells (RBCs), also called erythrocytes, make up the largest cell population. RBCs constitute an offensive weapon because they transport oxygen to various body parts and they break down carbon dioxide to a less toxic form. The RBCs in humans and all other mammals (except members of the Family Camelidae, such as the camel) are biconcave, circular discs without nuclei.

RBCs are produced in the red bone marrow of certain bones. As they develop, they produce massive quantities of hemoglobin, the oxygen-transporting pigment that contains iron, which in the oxygenated form gives blood its red color. Worn-out RBCs are broken down at the rate of 2 million cells per second in the liver and spleen by phagocytic WBCs. Some of the components of the RBCs are then recycled in order for the body to maintain a constant number of these cells in the blood.

In this exercise, you will have an opportunity to stain differentially and to observe blood cell characteristics on slides.

Definitions

Amoeboid. To make movements or changes in shape by means of protoplasmic flow.

Basophil. A granulocyte in which the cytoplasmic granules stain dark purplish blue with methylene blue, a blue basophilic-type dye found in Wright's stain.

Eosinophil. A granulocyte in which the cytoplasmic granules stain red with eosin, a red acidophilic-type dye found in Wright's stain.

Lymphocyte. A colorless agranulocyte produced in lymphoid tissue. It has a single nucleus and very little cytoplasm. They often look like dark purple round dots.

Megakaryocyte. A large cell with a lobulated nucleus that is found in bone marrow. It is the cell from which platelets originate.

Monocyte. A large agranulocyte normally found in the lymph nodes, spleen, bone marrow, and loose connective tissue. It is phagocytic with sluggish movements.

Neutrophil. A mature granulocyte present in peripheral circulation. The cytoplasmic granules stain poorly or not at all with Wright's stain. The nuclei of most neutrophils are large, contain several lobes, and are described as polymorphonuclear (PMN) leukocytes.

Platelet. A small oval-to-round, pink fragment, 3 microns in diameter. Plays a role in clotting blood.

Objectives

1. Describe historical and background information on blood and some of its microscopic cell types, their origin, morphology, number, and role in fighting disease.

2. Interpret stained blood slides: observe the cellular appearance of normal blood and perform a differential WBC count.

Prelab Questions

1. Name the three types of granulated white blood cells.

2. What cell is important for blood clotting?

3. Which blood cell is increased in a bacterial infection?

Materials

Either animal blood or prepared commercial unstained or stained human peripheral blood smears. For student use, blood should be dispensed with either a plastic dropper or a dropping bottle capable of dispensing a small drop.

Note: In the event of spilled blood, use disposable gloves and towels to remove blood. Then disinfect the area with a germicide, such as hypochlorite bleach diluted approximately 1:20 with water.

For use with whole blood

New microscope slides, 3

Plastic droppers for dispensing blood on slides, dispensing Wright's stain, and adding phosphate buffer

Hazardous waste container for droppers and slides

For use with whole blood and unstained prepared slides

Wright's stain (or Wright-Giemsa), dropping bottle, 1 per two students

Phosphate buffer, pH 6.8, dropping bottle, 1 per two students

Wash bottle containing distilled water, 1 per two students

A Coplin jar with 95% ethanol

Staining rack

Colored pencils for drawings: pink, blue, purple, or lavender

Procedure

We recommend using animal blood from a cow, pig, or dog or commercially prepared slides to minimize disease risk.

Safety Precautions: Working with blood carries risks from unknown contaminants. Disposable gloves and goggles must be worn when preparing slides.

1. Prepare three clean microscope slides free of oil and dust particles as follows:
   a. Wash slides with a detergent solution and then rinse thoroughly.
   b. Immerse slides in a jar of 95% alcohol.
   c. Air dry and polish with lens paper.
2. Place a drop of blood on one end of a clean slide (figure 26.2a). Repeat with a second clean slide.
3. Spread the drop of blood on the slide as follows:
   a. Place the slide on your laboratory bench-top. With your thumb and middle finger, firmly hold the sides of the slide on the end where the drop of blood is located.

b. With your other hand, place the narrow edge of a clean slide approximately ½" in *front* of the drop at an angle of about 30° (figure 26.2b).

c. Carefully push the spreader slide *back* until it comes in contact with the drop, at which point the drop will spread outward to both edges of the slide (figure 26.2c).

d. Immediately, and with a firm, steady movement, push the blood slowly toward the opposite end of the slide (figure 26.2d).

Note: Use of these procedural restraints (a small drop; a small spreader slide angle; and a *slow*, steady spreader slide movement) should provide a thin film for study of RBCs. A good smear has the following characteristics: smooth, without serrations; even edges; and uniform distribution over the middle two-thirds of the slide.

e. Allow slide to air dry for 5 minutes. Do not blot.

f. For the second slide, prepare a thicker film by using a larger spreader slide angle (45°) and by pushing the blood more rapidly to the opposite end of the slide. The second slide is best for determining the differential WBC count.

Note: The unused end of the first spreader slide can be used to prepare the second slide. Discard used spreader slides in the hazardous waste container.

4. Stain the blood smears with Wright's stain as follows:
   a. Suspend the slides so that they lie *flat* on the staining rack supports.
   b. Flood or add 15 drops of Wright's stain to each blood smear. Let it stand for 3–4 minutes. This fixes the blood film.
   c. Without removing the stain, add an equal volume of phosphate buffer. Blow gently through a pipet on each side of the slide to help ensure mixing of stain and buffer solutions.
   d. Let stand until a green, metallic scum forms on the surface of the slide (usually within 2–4 minutes).

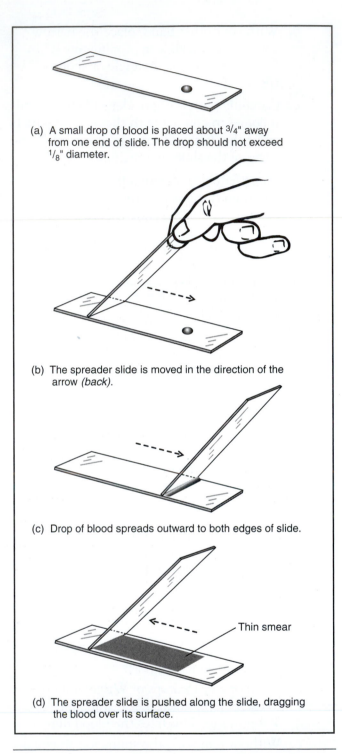

(a) A small drop of blood is placed about 3/4" away from one end of slide. The drop should not exceed 1/8" diameter.

(b) The spreader slide is moved in the direction of the arrow (back).

(c) Drop of blood spreads outward to both edges of slide.

Thin smear

(d) The spreader slide is pushed along the slide, dragging the blood over its surface.

Figure 26.2 Method for preparing a blood smear.

e. Wash off the stain with water. Begin washing while the stain is on the slide in order to prevent precipitation of scumlike precipitate, which cannot be removed.

The initial purple appearance on the slide should be washed until it is lavender-pink.

f. Wipe off excess stain from the back of the slide and allow it to stand on end to dry (which is preferable to drying between bibulous paper).

5. Examine stained blood smears:

a. Make an initial examination of the first blood smear with the low-power objective lens to find the most suitable areas for viewing with the oil immersion objective lens.

b. Next, using the oil immersion lens, make a study of the various WBC types present: basophils, eosinophils, lymphocytes, monocytes, neutrophils, and platelets. For help in this study, consult color plate 48, the Definitions section describing their staining characteristics with Wright's stain, and figure 26.1. Prepare color drawings of your findings in the Laboratory Report.

c. Conduct a differential WBC count using the second blood smear. For normal blood with a leukocyte count of 5,000–10,000 WBCs/ml, you would classify 100 leukocytes. In order to do this, you may have to examine the number and kinds of WBCs present in as many as twenty microscopic fields. Record your findings in table 26.2 of the Laboratory Report and calculate the percentage of each WBC type.

Table 26.1 Cellular Description of Normal Blood*

Total Counts	Differential WBC Counts
RBC 5,200,000/ml	Neutrophils 64%
WBC 7,200/ml	Lymphocytes 33%
Platelets 350,000/ml	Monocytes 2%
Basophils 50/ml	Eosinophils 1%

*From R. Kracke. *Diseases of the blood and atlas of hematology*, Second Edition. Copyright 1941. J. B. Lippincott Co. By permission.

EXERCISE

26

Laboratory Report: Differential White Blood Cell Stains

Results

1. Color drawings of RBCs and various WBCs found in blood smears stained with Wright's stain.

RBCs   Neutrophils   Eosinophils

Basophils   Monocytes   Lymphocytes

2. Differential WBC count. In table 26.2, record the kinds of leukocytes found as you examine each microscopic field. After counting 100 WBCs, calculate their percentages from the totals found for each type. Also record the number of microscopic fields examined to find 100 WBCs: _____.

Table 26.2 Kinds and Percentages of WBCs Found in Blood Smear

Neutrophils	Eosinophils	Basophils	Lymphocytes	Monocytes
Total				
Percentage				

Questions

1. What problems, if any, did you find in preparing and staining your blood smears? Indicate any differences noted between thin and thick smears.

2. Were your blood stains satisfactory? Did the stained cells resemble those in figure 26.1 and color plate 48? Were they better?

3. Did your differential white blood cell count percentages compare with the percentages in normal blood (table 26.1)? If not, give an explanation.

4. Were there any WBC types that you did not find in your blood smear? If so, which one(s)? Why did you not find them?

5. Matching (you may wish to consult your text).
   a. Neutrophils   _____ Involved in antibody production
   b. Basophils   _____ A phagocytic cell
   c. Monocytes   _____ Increased number in parasitic infections
   d. Eosinophils   _____ Turn into macrophages
   e. Lymphocytes   _____ Release histamine

27

Lysozyme, an Enzyme Contributing to Innate Immunity

Getting Started

A number of antimicrobial substances have been isolated from animal cells and body fluids. One of these is the protein **lysozyme,** discovered in 1922 by Alexander Fleming. This is an important enzyme contributing to **innate immunity** in animals. Unlike adaptive immunity, which requires an animal to be first exposed to the pathogen for protection, innate immunity protects the animal at birth.

Lysozyme is a widely distributed **enzyme** found not only in white blood cells, egg white, tears, saliva, human milk, and nasal secretions but also in insects and plants. It is able to enzymatically break the bonds of the **peptidoglycan layer** of the bacterial cell wall, thereby weakening and eventually destroying it. It is particularly active against Gram-positive bacteria that have an exposed peptidoglycan layer. Gram-negative cells have a membrane surrounding the peptidoglycan layer, which gives them protection, but if lysozyme can make contact with the inner wall, they too are killed. Cell walls of yeast do not contain peptidoglycan and usually are not affected by the same drugs that inhibit bacteria.

Penicillin also acts on the cell wall, but its action is quite different. It binds to proteins involved in cell wall synthesis rather than the cell wall itself. Therefore, penicillin is most effective on growing, dividing cells as it prevents cell wall formation. Lysozyme acts on the cell wall itself, and the cells do not need to be dividing for it to be effective.

Lysozyme is an important first line of defense. The eyes are moist and vulnerable to infection so lysozyme is valuable for preventing conjunctivitis, an infection of the eye. Children fed milk from a bottle rather than breast-fed have a higher incidence of illness because they lack the protection of this enzyme.

Lysozyme is also found in the tails of bacteriophage, where it expedites entry of the phage into the cell. Later phage stages direct the host cell to produce more lysozyme for use in facilitating release of new phage particles.

In this exercise, you will have an opportunity to assay the antimicrobial activity of lysozyme collected from tears and egg white and to compare this activity with a commercial lysozyme preparation of known activity.

Definitions

Innate immunity. An immunity to infectious disease in a species occurring as a part of its natural biological makeup.

Lysozyme. An enzyme able to break down and destroy bacterial cell walls. It occurs naturally in tears, saliva, phagocytic white blood cells, and egg white.

Peptidoglycan layer. The rigid backbone of the bacterial cell wall, composed of repeating subunits of N-acetylmuramic acid and N-acetylglucosamine and other amino acids.

Proteolytic enzyme. An enzyme able to hydrolyze proteins or peptides with the resulting formation of simpler, more soluble products such as amino acids.

Objectives

1. Describe innate immunity and how such chemicals as lysozyme function in immunity.
2. Assess antimicrobial activity of various natural lysozyme preparations.

Prelab Questions

1. How will you obtain tears for your experiment?
2. How will you be able to tell that the lysozyme worked against the microbe?
3. Besides the eyes, where else is lysozyme found?

Materials

Per pair of students

Trypticase soy broth cultures (37°C, 24-hour) of *Staphylococcus epidermidis* and *E. coli*

Sabouraud's dextrose broth culture (25°C, 24-hour) of *Saccharomyces cerevisiae*

Trypticase soy agar plates, 2

Sabouraud's dextrose agar plate, 1

Petri dish containing 9 sterile filter paper discs (approximately 1/2" diameter)

Petri dish containing 1–2 ml of aseptically prepared raw egg white

Petri dish containing 1–2 ml of lysozyme chloride (Sigma-Aldrich cat. # L-28791G) with activity of approximately 45,000 units per mg of protein; diluted 1:10 with sterile distilled water

Raw onion

Scalpel or sharp knife for cutting the onion

Mortar and pestle

Pair of tweezers

Ruler divided in mm

PROCEDURE

First Session

1. With a glass marking pen, label a trypticase soy agar plate on the bottom with your name and date and divide into thirds. Label one part T for tears, another part EW for egg white, and the remaining part L for lysozyme. Label it *Staphylococcus*.

2. Repeat, labeling a second trypticase soy agar plate with *E. coli* and the Sabouraud's plate with *Saccharomyces*.

3. Swab the surface of each plate with the appropriate culture. Dip a swab once in the broth culture and swab the plate in three directions to spread the microorganisms evenly around the plate.

4. In order to induce tears, one student should remove the outer skin from an onion. Then cut the onion into small pieces and crush in a mortar with a pestle. The other student should be prepared to collect the secreted tears in a sterile petri dish (0.5–1 ml is sufficient). Students do not need to contribute tears if they are uncomfortable with the procedure.

5. Sterilize tweezers by dipping in 95% alcohol, then touching the tweezers to a flame, permitting the alcohol to burn off. Never put the tweezers directly in the flame.

6. Select a sterile filter paper disc from a petri dish and dip into the tears solution. Transfer the moistened disc to the center of the area marked T on the inoculated agar plate. Alcohol sterilize the tweezers before picking up another paper disc.

7. Repeat step 6 transferring egg white and lysozyme solution to the appropriate agar plates.

8. Incubate plates inoculated with *Staphylococcus* and *Escherichia coli* inverted at 37°C for 24 hours and *Saccharomyces cerevisiae* at room temperature for 48 hours.

Second Session

1. Observe the petri dishes for zones of inhibition around the filter paper discs. With a ruler, determine their diameter in mm, and record your results in table 27.1 of the Laboratory Report.

EXERCISE

27

Laboratory Report: Lysozyme, an Enzyme Contributing to Innate Immunity

Results

1. Complete table 27.1 (see instructions in Procedure, session 2, step 1).

Table 27.1   Antimicrobial Activity of Various Lysozyme Extracts

Test Organism	Diameter of Inhibition Zone (mm)		
	Tears	Egg White	Lysozyme*
S. epidermidis			
E. coli			
S. cerevisiae			

*1:10 dilution

2. Which of the three preparations was the most active? _____ Least active? _____ Consider the lysozyme *dilution* when preparing your answer.

3. Which organism(s) were not inhibited by lysozyme?

4. Calculate a rough estimate of the lysozyme activity of tears and egg white compared to the pure lysozyme.

Table 27.2   Units of Lysozyme Activity for Egg White and Tears

Test Organism	Units of Lysozyme Activity/Mg of Protein	
	Egg White	Tears
S. epidermidis		
E. coli		

Questions

1. How are penicillin and lysozyme similar in their mode of action?

2. How are they different?

3. What large groups of microorganisms are susceptible to lysozyme? Resistant to lysozyme?

4. Why is lysozyme considered to contribute to innate immunity?

5. Why are most Gram-negative bacteria not lysed by lysozyme, yet they have a peptidoglycan cell wall structure similar to that of the Gram-positive bacteria?

28

Enzyme-Linked Immunosorbent Assay (ELISA)

Getting Started

The ELISA test is widely used for identification of plant, animal, and human pathogens, including viruses. It is used as an initial screening test for diseases such as hepatitis A, B, and C and AIDS. In the clinical setting, it is used to identify a variety of microbial pathogens because of test sensitivity and simplicity, often requiring only a swab or blood sample from the infected host. Of great importance to the success of this technique is the plastic microwell plate, which can attract the reactants to its surface and hold on to them. In this exercise, the ELISA technique is used to identify a sexually transmitted disease, HIV.

Various modifications of the ELISA antigen–antibody technique exist. In the direct immunosorbent assay, known as the double antibody sandwich assay (figure 28.1a), the microwell is coated with antibodies. In an indirect immunosorbent assay (figure 28.1b), the patient's antiserum is added to a microwell previously coated with antigens. If antibodies related to the antigens are present, they attach to the adsorbed antigens. After washing to remove unbound specimen components, an antibody conjugate that has coupled to the enzyme, horseradish peroxidase, is added. If binding occurs between the antigen and the antibody conjugate, a sandwich is formed, containing adsorbed antigens, patient antibodies, and the horseradish peroxidase enzyme. (Peroxidases are enzymes that catalyze the oxidation of organic substrates.) Next, the organic substrate used for this test, urea peroxide, is added. When oxidized by the peroxidase enzyme, free oxygen (O) is released. A color indicator, tetramethylbenzidine, is added, which when oxidized by the free oxygen, produces a yellow color (color plate 49). Lack of color means that the patient's antiserum does not contain antibodies to that disease. Once the reaction is complete, the color change is determined spectrophotometrically on a plate reader that calculates cutoff values for wells testing positive. Samples testing positive are repeated, and if they are still positive, a Western blot is performed to confirm the positive result. Specimens that yield an indeterminate result are retested in a clinical laboratory. If the retest result was still indeterminate, a second specimen would be obtained. Absorbances obtained with such reactions may be lower than expected but are still positive.

You will be performing a simulated ELISA for which you can visually determine the results.

Objectives

1. Explain the main steps in performing an ELISA test.
2. Interpret positive and negative ELISA test results.

Prelab Questions

1. What is the purpose of the chromogen?
2. What is attached to the microtiter well in an indirect ELISA?
3. What is the color of a positive ELISA test?

Materials

An AIDS Simutest Kit that contains the following (per group):

　Antigen-coated microtiter plate containing 96 microwells

　Serum from patients A through H (can be divided so each group does all samples or only a few (A–D) as time allows)

　Conjugate bottle

　Chromogen bottle

　Pipet(s) (the ones from the kit are calibrated)

Other materials (per group)

　Beaker (100 ml or larger)

　Distilled or deionized water (200–400 ml), distilled water bottles work well

　Timer

　Gloves (recommended)

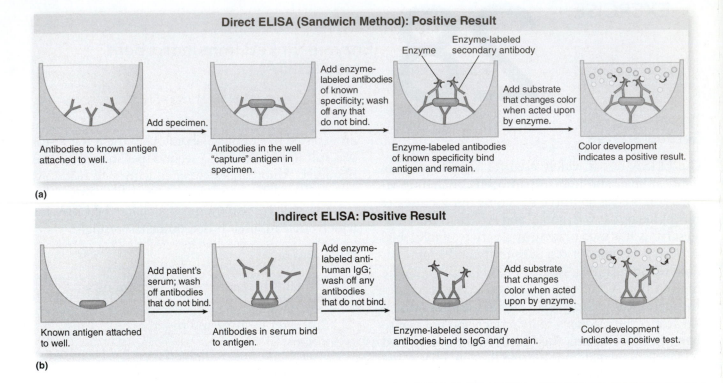

Figure 28.1 The ELISA technique. (*a*) The direct sandwich method, and (*b*) The indirect antibody method.

PROCEDURE

Note: This is a **SIMULATED** HIV test, and no infectious materials are present in this kit. Due to some of the chemicals used in this test, gloves are recommended but not required.

1. Each group should obtain a microtiter plate, the serum samples assigned, a conjugate bottle, a chromogen bottle, a pipet, a beaker with distilled water, and extra distilled water.

2. Using the pipet, add 6 drops of serum sample to the first two wells of the row. Sample A should be put in row A.

3. Using the water in the beaker, rinse out the pipet 4 or 5 times.

4. Add 6 drops of serum sample B to the first two wells of row B. Rinse the pipet out.

5. Repeat these steps until you have added each sample to the corresponding row.

6. Pour out the beaker water into the sink and add about 100 ml of fresh water to it.

7. Add 6 drops of distilled water to each well from wells 2–7 for each sample (A, B, etc.). **Leave wells 1 and 8 blank of distilled water.**

8. You will now be performing a serial dilution on the sample wells.

9. Mix well 2 of sample A by gently pipetting up and down. Take 6 drops from this second well, and transfer it to well 3. (If you removed too much liquid, add the 6 drops to well 3 and then add the rest back to the second well.) Mix well 3 and transfer 6 drops into well 4. Continue this serial dilution until well 8. (6 drops from #4 into #5, 6 drops from #5 into #6, 6 drops from #6 into #7, and 6 drops from #7 into #8, and discard 6 drops from #8).

10. Rinse the pipet out 4 or 5 times in the beaker water.

11. Repeat the serial dilution steps from step 9 for sample B, sample C, etc.

12. Allow the microtiter plate to incubate 10 minutes at room temperature.

13. Pour out the beaker water into the sink and add about 100 ml of fresh water to it.

14. Add 2 drops of conjugate to each well (1–7) in each row. Allow to incubate at room temperature for 5 minutes. (You do not need to pipet up and down.)

15. Rinse the pipet out.

16. Add 3 drops of chromogen into each well (1–7) in each row. (You do not need to pipet up and down.) A light yellow/colorless is considered to be negative. An orange/red color is considered positive.

17. Record your results in the table in the Laboratory Report.

18. Once you are finished recording your results, take your microtiter plate to the sink and **gently** wash out the wells. You do not want to dislodge the pellet at the bottom of the wells. Dry the bottom of the plate with a paper towel and leave the plates upside down to dry.

19. A negative result indicates that serum antibody to HIV antigens is either absent or below the level of detection of the assay, or the specimen was obtained too early in the response.

20. A positive result implies the presence of antibody to HIV. An early acute-phase patient may only present an IgM response, whereas the chronic or convalescent patient may only present an IgG response. Some test kits will only detect IgG or IgM, whereas others will test for both.

EXERCISE

28

Laboratory Report: Enzyme-Linked Immunosorbent Assay (ELISA)

Results and Questions

1. Discuss the test results and their significance.

	Well Color	Result	What does the result mean?
Sample A			
Sample B			
Sample C			
Sample D			
Sample E			
Sample F			
Sample G			
Sample H			

2. What is the importance of rinsing the wells when conducting a real (not simulated) ELISA test?

3. What is the purpose of the conjugate?

4. Discuss the pros and cons of using the indirect enzyme-linked immunosorbent assay (ELISA) and the double antibody sandwich ELISA assay. You may need to consult your text for the answer to this question.

INTRODUCTION to the Prevention and Control of Communicable Diseases

Communicable, or infectious, diseases are transmitted from one person to another. Transmission is either by direct contact with a previously infected person—for example, by touching—or by indirect contact with a previously infected person who has contaminated the surrounding environment.

A classic example of indirect contact transmission is an epidemic of cholera that occurred in 1854 in London. During a 10-day period, more than 500 people became ill with cholera and subsequently died. As the epidemic continued, John Snow and John York began a study of the area and were able to prove by **epidemiological methods** only (the bacteriological nature of illness was not known at that time) that the outbreak stemmed from a community well on Broad Street known as the Broad Street Pump* (introduction figure I.29.1). These researchers then discovered that sewage from the cesspool of a nearby home was the pollution source and that an undiagnosed intestinal disorder had occurred in the home shortly before the cholera outbreak. They also learned that neighboring people who abstained from drinking the pump water remained well, whereas many of those who drank the water succumbed to cholera.

Today the incidence of cholera, typhoid fever, and other infectious diseases rarely reaches **epidemic** proportions in those countries that have developed standards and regulations for control of environmental reservoirs of infection, particularly the major reservoirs: water, food, and sewage. (The subject of public health sanitation is presented in exercise 29, which discusses water microbiology.)

All public health sanitation studies come under the surveillance of public health agencies responsible for prevention and control of communicable diseases. Among these are the federal cabinet-level Department of Health and Human Services, which conducts preventive medicine research, provides hospital facilities for military personnel, and gives financial assistance to state and local health departments, as well as assistance at times to developing countries. Additionally, the Centers for Disease Control (CDC) in Atlanta plays an important role in the prevention and control of communicable diseases. All state and local government agencies also perform important public health services.

Perhaps the most important international health agency is the World Health Organization (WHO), headquartered in Geneva, Switzerland. WHO distributes technical information, standardizes drugs, and develops international regulations important for the control and eradication of epidemic diseases. For example, smallpox, which was once a widespread disease, is nonexistent today. Its current emphasis is on preventing HIV disease in mothers and children.

Finally, there are voluntary health organizations that help in some of the causes previously mentioned. A notable example is the Bill and Melinda Gates Foundation, which focuses on fighting diseases in developing countries.

Definitions

Epidemic. The occurrence in a community or region of a group of illnesses of similar nature, clearly in excess of normal expectancy.

Epidemiological methods. Methods concerned with the extent and types of illnesses and injuries in groups of people and with the factors that influence their distribution. This implies that disease is not randomly distributed throughout a population but rather that subgroups within a given population differ in the frequency of susceptibility to different diseases.

*Snow, John. "The Broad Street Pump," in Roueche, Berton (ed.), *Curiosities of medicine*. New York: Berkely, Medallion, ed., 1964.

(a)

(b)

Figure I.29.1 (a) The John Snow pub in London where epidemiologists go to celebrate the heroics of John Snow's early epidemiological efforts to stem a cholera epidemic. (b) A replica of the pump, with pump handle attached, a monument dedicated to Dr. Snow in July 1992. At the time of the epidemic, he was so convinced that the disease was being carried by water from the pump that he had the pump handle removed. Koch isolated and identified the cholera vibrio about 30 years later. Courtesy of Kathryn Foxhall.

EXERCISE

29

Bacteriological Examination of Water: Multiple-Tube Fermentation and Membrane Filter Techniques

Getting Started

Water, water, everywhere
Nor any drop to drink.

The Ancient Mariner, Coleridge, 1796

This rhyme refers to seawater, undrinkable because of its high salt content. Today the same can be said of freshwater supplies, polluted primarily by humans and their activities. A typhoid epidemic, dead fish on the beach, and red tides are all visible evidence of pollution. There are two primary causes of water pollution: dumping of untreated (raw) sewage and inorganic and organic industrial wastes, and fecal pollution by humans and animals of both freshwater and groundwater. In the United States, sewage and chemical wastes have been reduced largely as a result of federal and local legislation requiring a minimum of secondary treatment for sewage, as well as severe penalties for careless dumping of chemical wastes.

Fecal pollution is more difficult to control, particularly as the supply of water throughout the world becomes more critical. In developing countries, it is estimated that over 12,000 children die every day from diseases caused by waterborne fecal pollutants, such as cholera, typhoid fever, bacterial and amoebic dysentery, and viral diseases like polio and infectious hepatitis. Most inhabitants of these countries are in intimate contact with polluted water because not only do they drink it, but they also bathe, swim, and wash their clothes in it.

The increased organic matter in such water allows **anaerobic bacteria** to increase their numbers in relation to the **aerobic bacteria** originally present (figure 29.1). *Sphaerotilus natans* is a nuisance bacterium that forms an external sheath (figure 29.2), that allows it to adhere to water pipes and reduce their carrying capacity.

Microbes are also beneficial in water purification. In smaller sewage treatment plants, raw sewage is passed through a slow sand filter, wherein micro-organisms present in the sand degrade (metabolize)

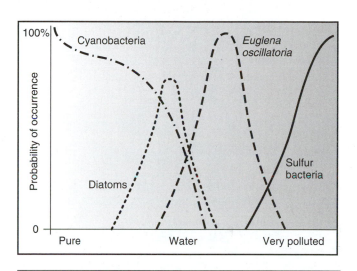

Figure 29.1 Diversity of microbes found in pure to very polluted water. Note the change from aerobic to anaerobic microbes as the water becomes more polluted.
Courtesy of Settlemire and Hughes, *Microbiology for Health Students*, Reston Publishing Co., Reston, Virginia.

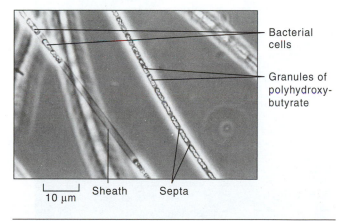

Figure 29.2 *Sphaerotilus* species, a sheathed bacterium that often produces masses of brownish scum beneath the surface of polluted streams.
Phase-contrast photomicrograph courtesy of J. T. Staley and J. P. Dalmasso.

organic waste compounds before the effluent is discharged. Sewage effluent is finally chlorinated to further reduce fecal microbial contaminants. The development of sewage treatment plants and the control of raw sewage discharge reduced the annual typhoid fever death rate in the United States from about 70 deaths per 100,000 population to nearly zero. However, the potential danger of pollution is always present. In 1975, residents of the city of Camas, Washington, were inundated by intestinal disorders that were traced to fecal pollution of the water supply by beavers infected with the protozoan *Giardia lamblia* (figure 29.3 and color plate 16).

Two microbiological methods are commonly used to determine if a given water sample is polluted:

1. *Total number of microorganisms present in the water.* The plate count method provides an indication of the amount of organic matter present. Due to the diversity in microbial physiology, no single growth medium and no single set of cultural conditions can satisfy universal microbial growth. Standard plate counts on nutrient agar incubated at two temperatures, 20°C and 35°C, provide a useful indication of the organic pollution load in water.

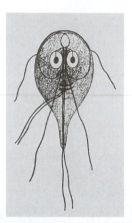

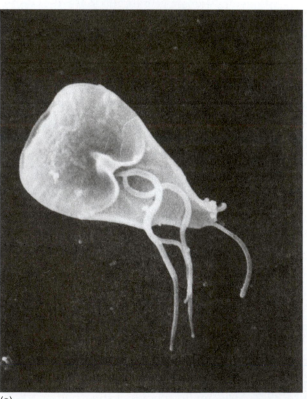

(a)

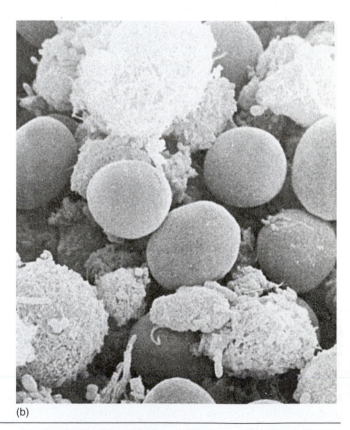

(b)

Figure 29.3 Human fecal specimen illustrations showing forms of *Giardia lamblia*, a waterborne protozoan pathogen that inhabits the intestinal tract of certain warm-blooded animals. (*a*) An electron microscope view of *Giardia lamblia*. (*b*) Smooth ovoid cysts. The cyst form is resistant to adverse environments and is the form released with fecal material. Scanning electron microscopy, magnification ×2,900, 1 micron = 2.9 mm. Cysts are embedded in a mat of debris, bacteria, and fecal material.

(*a*) Courtesy of the CDC. (*b*) From D. W. Luchtel, W. P. Lawrence, and F. B. De Walle, "Electron Microscopy of *Giardia lamblia* Cysts," *Applied and Environmental Microbiology*, 40:821–832, 1980. © American Society for Microbiology.

Newer methods of determining the total number of microorganisms include quantitative polymerase chain reaction (qPCR) and real-time PCR. These methods use fluorescent dyes to determine the presence and/or quantity of coliforms, so the target DNA sequence of the microbe(s) must be known. A disadvantage to this method currently is the cost.

2. *Indicator organisms.* Indicator organisms are normally nonpathogenic, always occur in large quantities in feces, and are relatively easy to detect compared to waterborne pathogens. Ten enzyme-based tests have been approved since 2002 by the Environmental Protection Agency (EPA) for detection of coliforms. One such test is the Colilert test which uses ONPG to detect total coliforms and turns yellow (exercise 18) and MUG (4-methylumbelliferyl-β-D-glucuronide), which fluoresces blue, to detect *E. coli*. The diagnosis of pathogens is complicated and time-consuming, and thus less suited for routine investigations. Assuming pathogens are dying off faster than the indicator organisms, the absence of indicator organisms guarantees, in most cases, the absence of pathogens. In certain cases, newer techniques have shown the pathogens' increased resistance to the aqueous environment. Similar findings have been found with respect to chlorination resistance. Such results suggest that some of the principles and methods in conventional water examination are of questionable value.

Despite their limitations, conventional methods for detection of fecal contamination are still widely employed. In general, these procedures are the same ones described by the American Public Health Association in *Standard Methods for the Examination of Water and Wastewater.*

The indicator organisms most widely used are **coliform bacteria.** These include all aerobic and facultative anaerobic, Gram-negative, non-spore-forming, rod-shaped bacteria that ferment lactose with gas formation within 48 hours at 37°C, and comprise *Escherichia coli* (6^{10} to 9^{10} cells/g of feces) together with a number of closely related organisms (exercise 24). Noncoliforms that are sometimes employed, primarily for confirmation, include *Streptococcus faecalis*, some related species, and, in Great Britain, *Clostridium perfringens* (also called *Clostridium welchii*).

The presence of *E. coli* in water from such sources as reservoirs suggests that chlorination is inadequate. Current standards for drinking water state that it should be free of coliforms and contain no more than ten other microorganisms per ml.

Two of the most important methods applied to detect coliform organisms are the multiple-tube fermentation technique and the membrane filter technique.

Multiple-Tube Fermentation Technique

This technique employs three consecutive tests: first, a presumptive test; if the first test is positive, then a confirmed test; and, finally, a completed test (figure 29.4).

Presumptive Test A specific enrichment procedure for **coliform bacteria** is conducted in fermentation tubes filled with a **selective growth medium** (lauryl lactose tryptose broth), which contain inverted Durham tubes for detection of fermentation gas (figure 29.4).

The selective factors found in the medium are lactose and sodium lauryl sulfate; often, a pH indicator dye for detecting acid production, such as bromcresol purple or basic fuchsin, is added. Many bacteria cannot ferment lactose, and sodium lauryl sulfate inhibits many other bacteria, including spore formers, thus making it selective.

Tube turbidity ensures bacterial presence. The formation of gas in the Durham tube within 24–48 hours constitutes a positive presumptive test for coliforms and the possibility of fecal pollution. The test is presumptive only; several other bacteria produce similar results.

The most probable number test (MPN) is useful for counting bacteria that reluctantly form colonies on agar plates or membrane filters but grow readily in liquid media. In principle, the water sample is diluted so some broth tubes contain one bacterial cell. After incubation, some broth tubes show growth with gas, whereas others do not. The total viable count is then determined by counting the portion of positive tubes and referring to a statistical MPN table used for calculating the total viable bacterial count (table 29.1).

Confirmed Test This test confirms the presence of coliform bacteria when either a positive or doubtful presumptive test is obtained. A loopful of growth from a presumptive tube is transferred into a tube of 2% brilliant green bile lactose broth

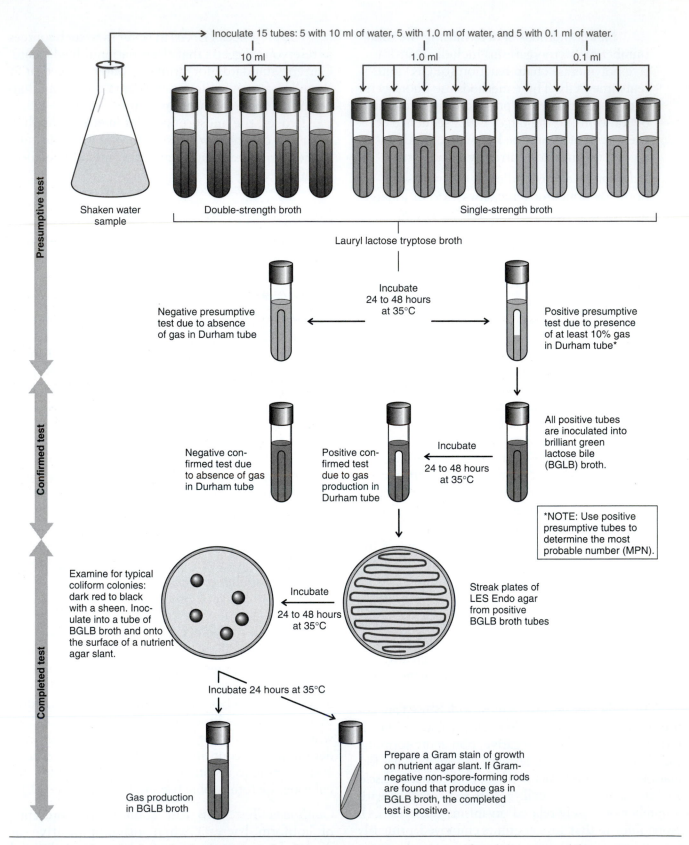

Inoculate 15 tubes: 5 with 10 ml of water, 5 with 1.0 ml of water, and 5 with 0.1 ml of water.

10 ml   1.0 ml   0.1 ml

Shaken water sample

Double-strength broth   Single-strength broth

Lauryl lactose tryptose broth

Presumptive test

Incubate 24 to 48 hours at 35°C

Negative presumptive test due to absence of gas in Durham tube

Positive presumptive test due to presence of at least 10% gas in Durham tube*

Confirmed test

All positive tubes are inoculated into brilliant green lactose bile (BGLB) broth.

Negative confirmed test due to absence of gas in Durham tube

Positive confirmed test due to gas production in Durham tube

Incubate 24 to 48 hours at 35°C

*NOTE: Use positive presumptive tubes to determine the most probable number (MPN).

Completed test

Examine for typical coliform colonies: dark red to black with a sheen. Inoculate into a tube of BGLB broth and onto the surface of a nutrient agar slant.

Incubate 24 to 48 hours at 35°C

Streak plates of LES Endo agar from positive BGLB broth tubes

Incubate 24 hours at 35°C

Gas production in BGLB broth

Prepare a Gram stain of growth on nutrient agar slant. If Gram-negative non-spore-forming rods are found that produce gas in BGLB broth, the completed test is positive.

Figure 29.4 Standard methods procedure for the examination of water and wastewater and for use in determining most probable number (MPN).

containing a Durham tube and incubated at 35°C for 48 hours. This selective medium detects coliform bacteria in water and dairy food products. The correct concentration of the dye (brilliant green) and bile must be present. If the concentration is too high, coliform growth can also be inhibited. Bile is naturally found in the intestine, where it serves a similar purpose—encouraging growth of coliform bacteria while discouraging growth of other bacteria. The presence of gas in the Durham tube after incubating for 24–48 hours constitutes a positive confirmed test.

Completed Test This test helps to further confirm doubtful and, if desired, positive confirmed test results. The test has two parts:

1. A plate of LES (Lawrence Experimental Station) Endo agar is streaked with a loopful of growth from a positive confirmed tube and incubated at 35°C for 24 hours. Typical coliform bacteria (*Escherichia coli* and *Enterobacter aerogenes*) exhibit good growth on this medium and form red to black colonies with a sheen. *Salmonella typhi* exhibits good growth, but the colonies are colorless. *Staphylococcus* growth is inhibited.

2. Next, a typical coliform colony from an LES Endo agar plate is inoculated into a tube of brilliant green bile broth and on the surface of a nutrient agar slant. They are then incubated at 35°C for 24 hours. A Gram stain is then prepared from growth present on the nutrient agar slant. The presence of gas in the brilliant green bile broth tube and Gram-negative, non-spore-forming rods constitutes a positive completed test for the presence of coliform bacteria; this infers possible contamination of the water sample with fecal matter.

Membrane Filter Technique

For this technique, a known volume of water sample (100 ml) is filtered by suction through a sterile polycarbonate or nitrocellulose acetate membrane filter. The filter is very thin (150 μm), and has a pore diameter of 0.45 μm; bacteria larger than 0.47 μm cannot pass through it. Filters with printed grid patterns facilitate colony counting. Sample turbidity presents a serious obstacle if suspended matter clogs

the filter before the required volume of water has passed through. Unknown samples must be diluted.

Once the water is filtered, the filter disc is aseptically transferred to the surface of a wetted pad soaked with Endo broth contained in a petri dish and incubated at 35°C for 24 hours. Coliform colonies appear pink to dark red in color with a golden metallic sheen. The number of characteristic coliform colonies is counted and the total number of coliform bacteria present in the original water sample can be calculated. The membrane filter method provides accurate results if the coliform colony count is 30–300 organisms per filter disc. For additional confirmation, the procedure for the completed multiple-tube fermentation test can be applied.

The advantages of the membrane filter technique over the multiple-tube fermentation test are (1) better reproducibility of results; (2) greater sensitivity, because larger amounts of water can be used; and (3) shorter time (one-quarter the time) for obtaining results. This method has been recognized by the United States Public Health Service for detection of coliforms in water.

Definitions

Aerobic bacteria. Microbes that grow and multiply in the presence of free gaseous oxygen.

Anaerobic bacteria. Microbes that grow best or exclusively in the absence of free oxygen.

Coliform bacteria. A collective term for bacteria that inhabit the colon, are Gram-negative, and ferment lactose (page 239).

Indicator organism. A nonpathogenic organism whose presence in water or sewage serves as a sign of possible pollution with pathogens.

Selective growth medium. A growth medium containing substances that inhibit the growth of certain organisms but not others.

Objectives

1. Explain the use of a multiple-tube fermentation technique for detecting the presence and number of coliform indicator organisms present in water samples.

2. Explain the membrane filter technique for detecting the presence and number of coliform bacteria in water samples.

Prelab Questions

1. In the United States, which bacterium is used as an indicator organism for coliforms?

2. Name an advantage of the filter membrane technique over the multiple-tube fermentation test.

3. What color will a colony testing positive on the Endo media exhibit?

Tube containing 10 ml of sterile Endo broth (LES or MF), 1

Sterile 90 ml water blanks, 2

Erlenmeyer flasks containing 25 ml of sterile water, 6

Sterile 10 ml pipets, 2

Vacuum pump or Venturi vacuum system

Water sample for coliform analysis, 100 ml

Materials

Multiple-Tube Fermentation Technique
Per student (figure 29.4)

Test tubes (50 ml), containing 10 ml of double-strength lauryl sulfate (lauryl lactose tryptose) broth plus Durham tubes, 5

Small test tubes containing 10 ml of single-strength lauryl sulfate broth plus Durham tubes, 10

Sterile 10 ml pipet, 1

Sterile 1 ml pipet calibrated in 0.1 ml units, 1

Brilliant green bile 2% broth plus Durham tubes, 2

LES Endo agar plate, 1

Nutrient agar slant, 1

Sterile 100 ml screw-cap bottle for collecting water sample, 1

Membrane Filter Technique
Demonstration (figure 29.5)

A 1-liter side-arm Erlenmeyer flask, 1

Sterile membrane filter holder assembly, two parts wrapped separately (figure 29.5, frames 2 and 3), 1 unit

A metal clamp for clamping filter funnel to filter base

Sterile membrane filters, 47 mm diameter, 0.45 μm pore size

Forceps, 1 pair

Sterile 50 mm diameter petri dishes, 3

Absorbent filter pads, 3

PROCEDURE

Multiple-Tube Fermentation Technique

Note: Your instructor may ask you to bring a 50–100 ml sample of water from home, a nearby stream, a lake, or some other location for analysis. When taking a tap sample, the orifice of the water tap should be flamed before being opened and water should run for 5 to 10 minutes with the tap in the same position to prevent loosening of bacteria from inside the tap. Next, using aseptic technique, open a sterile bottle (obtained from the instructor beforehand) and collect a sample. If the sample cannot be examined within 1–2 hours, keep refrigerated until ready to test.

First Session (Presumptive Test)

1. Shake the water sample. Aseptically pipet 10 ml of the sample into each of the five large tubes containing 10 ml aliquots of double-strength lauryl sulfate broth. Next, with a 1 ml pipet, transfer 1 ml of the water sample into five of the smaller tubes and then 0.1 ml into the remaining five small tubes of lauryl sulfate broth. Be sure to label the tubes.

2. Incubate the test tubes for 24–48 hours at 37°C.

Second and Third Sessions (Presumptive and Confirmed Tests)

1. After 24 hours, observe the tubes for gas production by gently shaking the tubes. If, after shaking, gas is not evident in the Durham tube, reincubate the tube for an additional 24 hours. Record any positive results for gas production in table 29.2 of the Laboratory Report.

Figure 29.5 Analysis of water for fecal contamination. Cellulose acetate membrane filter technique.
(1) Sterile membrane filter (0.45 μm) with grid for counting is handled with sterile forceps. *(2)* The sterile membrane filter is placed on filter holder base with grid side up. *(3)* The apparatus is assembled. *(4)* Sterile absorbent pads are aseptically placed in the bottom of three sterile petri dishes. *(5)* Each absorbent pad is saturated with 2.0 ml of Endo MF broth. *(6)* A portion of well-mixed water sample is poured into assembled funnel and filtered by vacuum. *(7)* Membrane filter is carefully removed with sterile forceps after disassembling the funnel. *(8)* Membrane filter is centered on the surface of the Endo-soaked absorbent pad (grid side up) using a slight rolling motion. *(9)* After incubation, the number of colonies on the filter is counted. The number of colonies on the filter reflects the number of coliform bacteria present in the original sample.

2. Observe the tubes for gas production and turbidity after 48 hours of incubation. If neither gas nor turbidity is present in any of the tubes, the test is negative. If turbidity is present but gas is not, the test is questionable and record as negative. If the tube shows gas production, the test is positive for coliform bacteria. Record your results in table 29.2 of the Laboratory Report.

3. MPN determination. Using your fermentation gas results in table 29.2, determine the number of tubes from each set containing gas. Determine the MPN by consulting table 29.1; for example, if you have gas in two of the first five tubes, in two of the second five tubes, and none in the third three tubes, your test readout is 2-2-0. Table 29.1 shows that the MPN for this readout is 9. Thus, your water sample contains nine organisms per 100 ml water with a 95% statistical probability of there being between three and twenty-five organisms.

Note: If your readout is 0-0-0, it means that the MPN is less than two organisms per 100 ml of water. Also, if the readout is 5-5-5, it means the MPN is greater than 1,600 organisms/100 ml water. In this instance, what procedural modification is required to obtain a more significant result? Report your answer in question 10 of the Questions section in the Laboratory Report.

4. The confirmed test should be administered to all tubes demonstrating either a positive or doubtful presumptive test. Inoculate a loopful of growth from each tube showing gas or dense turbidity into a tube of 2% brilliant green lactose bile broth. Incubate the tube(s) at 37°C for 24–48 hours.

Note: For expediency, your instructor may instruct you to inoculate only one tube. If so, for the inoculum use the tube of lauryl sulfate broth testing positive with the least inoculum of water.

Fourth Session (Confirmed and Completed Tests)

1. Examine the 2% brilliant green lactose bile tube(s) for gas production. Record your findings in the confirmed test section of the Laboratory Report.

2. Streak a loopful of growth from a positive tube of 2% brilliant green lactose bile broth on the surface of a plate containing LES Endo agar. Incubate at 37°C for 24 hours.

Fifth Session (Completed Test)

1. Examine the LES Endo agar plate(s) for the presence of typical coliform colonies (dark red to black with a sheen). Record your findings in the completed test section of the Laboratory Report.

2. With a loop, streak a nutrient agar slant with growth obtained from a typical coliform colony found on the LES Endo agar plate. Also inoculate a tube of 2% brilliant green lactose bile broth with growth from the same colony. Incubate the tubes at 37°C for 24 hours.

Sixth Session (Completed Test)

1. Examine the 2% brilliant green lactose bile broth tube for gas production. Record your result in the completed test section of the Laboratory Report.

2. Prepare a Gram stain of some of the growth present on the nutrient agar slant. Examine the slide for the presence of Gram-negative, non-spore-forming rods. Record your results in the completed test section of the Laboratory Report. The presence of gas and of Gram-negative, non-spore-forming rods constitutes a positive completed coliform test.

Membrane Filter Technique

First Session

1. Shake the water sample. Prepare two dilutions by transferring successive 10 ml aliquots into 90 ml blanks of sterile water (10^{-1} and 10^{-2} dilutions).

Note: Reshake prepared dilutions before using.

2. Assemble the filter holder apparatus as follows (figure 29.5):

   a. Using aseptic technique, unwrap the lower portion of the filter base, and insert a rubber stopper.

   b. Insert the base in the neck of the 1-liter side-arm Erlenmeyer flask (figure 29.5, frame 2).

Table 29.1 MPN Index and 95% Confidence Limits for Various Combinations of Positive Results When Five Tubes Are Used per Dilution (10 ml, 1.0 ml, 0.1 ml)

Combination of Positives	MPN Index/ 100 mL	95% Confidence Limits		Combination of Positives	MPN Index/ 100 mL	95% Confidence Limits	
		Lower	Upper			Lower	Upper
0-0-0	<2	—	—	4-3-0	27	12	67
0-0-1	3	1.0	10	4-3-1	33	15	77
0-1-0	3	1.0	10	4-4-0	34	16	80
0-2-0	4	1.0	13	5-0-0	23	9.0	86
1-0-0	2	1.0	11	5-0-1	30	10	110
1-0-1	4	1.0	15	5-0-2	40	20	140
1-1-0	4	1.0	15	5-1-0	30	10	120
1-1-1	6	2.0	18	5-1-1	50	10	150
1-2-0	6	2.0	18	5-1-2	60	30	180
2-0-0	4	1.0	17	5-2-0	50	20	170
2-0-1	7	2.0	20	5-2-1	70	30	210
2-1-0	7	2.0	21	5-2-2	90	40	250
2-1-1	9	3.0	24	5-3-0	80	30	250
2-2-0	9	3.0	25	5-3-1	110	40	300
2-3-0	12	5.0	29	5-3-2	140	60	360
3-0-0	8	3.0	24	5-3-3	170	80	410
3-0-1	11	4.0	29	5-4-0	130	50	390
3-1-0	11	4.0	29	5-4-1	170	70	480
3-1-1	14	6.0	35	5-4-2	220	100	580
3-2-0	14	6.0	35	5-4-3	280	120	690
3-2-3	17	7.0	40	5-4-4	350	160	820
4-0-0	13	5.0	38	5-5-0	240	100	940
4-0-1	17	7.0	45	5-5-1	300	100	1300
4-1-0	17	7.0	46	5-5-2	500	200	2000
4-1-1	21	9.0	55	5-5-3	900	300	2900
4-1-2	26	12	63	5-5-4	1600	600	5300
4-2-0	22	9.0	56	5-5-5	≥1600	—	—
4-2-1	26	12	65				

From *Standard Methods for the Examination of Water and Wastewater*, 18th edition. Copyright 1992 by the American Public Health Association.

c. With sterile forceps (sterilize by dipping in alcohol and flaming them with the Bunsen burner), transfer a sterile membrane filter onto the glass or plastic surface of the filter holder base (figure 29.5, frames 1 and 2). Make certain the membrane filter is placed with the ruled side up.

d. Aseptically remove the covered filter funnel from the butcher paper, and place the lower surface on top of the membrane filter. Clamp the filter funnel to the filter base with the clamp provided with the filter holder assembly (figure 29.5, frame 3).

3. Prepare 3 plates of Endo medium by adding 2 ml of the tubed broth to sterile absorbent pads previously placed aseptically (using sterile tweezers) on the bottom of the three petri dishes (figure 29.5, frames 4 and 5).

4. Remove the aluminum filter cover and pour the highest water dilution (10^{-2}) into the funnel (figure 29.5, frame 6). Assist the filtration process by turning on the vacuum pump or Venturi vacuum system.

5. Rinse the funnel walls with two 25 ml aliquots of sterile water to recover any bacteria that adhered to the wall of the funnel. This ensures accurate results.

6. Turn off (break) the vacuum and remove the filter holder funnel. Using aseptic technique and sterile tweezers, transfer the filter immediately to the previously prepared petri dish (figure 29.5, frames 7 and 8). Using a slight rolling motion, center the filter, grid side up, on the medium-soaked absorbent pad. Take care not to trap air under the filter, as this will prevent nutrient media from reaching all of the membrane surface (figure 29.5, frame 9).

7. Reassemble the filter apparatus with a new membrane filter, and repeat the filtration process first with the 10^{-1} water sample and finally with a 100 ml aliquot of the undiluted water sample.

8. Label and invert the petri dishes to prevent condensation from falling on the filter surface during incubation. Incubate plates for 24 hours at 37°C.

Second Session

1. Count the number of coliform bacteria by using either the low-power objective lens of the microscope or a dissecting microscope. Count only those colonies that exhibit a pink to dark-red center with or without a distinct golden metallic sheen.

2. Record the number of colonies found in each of the three dilutions in table 29.3 of the Laboratory Report.

EXERCISE

29

Laboratory Report: Bacteriological Examination of Water: Multiple-Tube Fermentation and Membrane Filter Techniques

Results

1. Multiple-Tube Fermentation Technique
   a. Record the results of the presumptive test in table 29.2.

Table 29.2 Presumptive Test for the Presence or Absence of Gas and Turbidity in Multiple-Tube Fermentation Media

Water Sample Size (ml)	Presence of Gas and Turbidity*				
	Tube #1	Tube #2	Tube #3	Tube #4	Tube #5
10					
1					
0.1					

*Use a + sign to indicate gas and a circle (O) around the plus sign to indicate turbidity.

   b. Determine the MPN below:
      Test readout: _____ MPN: _____ 95% Confidence Limits: _____
   c. Confirmed test results (gas production in brilliant green lactose bile 2% broth):
      Sample Number   ***Gas (+ or –)***
                          24 hours   48 hours

   d. Appearance of colonies on LES Endo agar:

   e. Completed test results:
      Sample Number   ***Gas (+ or –)***   **Gram-Stain Reaction**

2. Membrane Filter Technique

Table 29.3 Number of Coliform Colonies Present in Various Dilutions of the Water Sample

Undiluted Sample	10^{-1} Dilution	10^{-2} Dilution

a. Calculate the number of coliform colonies/ml present in the original water sample (show your calculations):

Questions

1. What nutritional means could be used to speed up the growth of the coliform organisms when using the membrane filter technique?

2. Describe two other applications of the membrane filter technique.

3. Why not test for pathogens such as *Salmonella* directly rather than using an indicator organism, such as coliform bacteria?

4. Why does a positive presumptive test not necessarily indicate that the water is unsafe for drinking?

5. List three organisms that usually give a positive presumptive test.

6. Describe the purpose of lactose and LES Endo agar in these tests.

7. What are some limitations of the membrane filter technique?

8. Define the term *coliform*.

9. Briefly explain what is meant by presumptive, confirmed, and completed tests in water analysis.

10. See Note 3, MPN determination, on page 244 for the question.

30

Epidemiology

Getting Started

Epidemiology is the study of factors that influence the spread and control of diseases within populations. Diseases such as colds, measles, and tuberculosis may be **communicable,** or **contagious,** when they are spread from one person to another. Communicable diseases are spread by either direct or indirect methods of transmission. Direct transmission occurs when individuals have person-to-person contact, such as with the transmission of fecal organisms from touching unwashed hands during handshaking or skin contact with a lesion generated from the herpes virus. Indirect transmission occurs when individuals transfer microorganisms to another person via respiratory droplets, when coughing or sneezing, or on **fomites,** inanimate objects such as doorknobs, pens, and toys that can harbor microorganisms.

In this lab, we will simulate the transmission of a "disease" from one person to another and demonstrate how an epidemiologist would track the spread of a communicable disease back to the original source, or **index case.** Using a piece of candy to transfer the "disease" agent to gloved hands, you will shake hands with classmates and transfer the "disease" to others. Contact with the "disease" agent will be determined by touching gloved hands (which had contact with the candy) to the surface of a nutrient agar plate. The presence of red colonies on the surface of the agar plate after 24 hours of incubation indicates contact with the "disease" agent, *Serratia marcescens.*

Definitions

Communicable. An infectious disease that is acquired either directly or indirectly from another host.

Contagious. A communicable disease that is easily transmitted from one person to another.

Fomites. Inanimate objects that can act as transmitters of pathogenic microorganisms or viruses.

Index case. The first identified case in an outbreak.

Objectives

1. Describe the common methods of communicable disease transmission.
2. Interpret data that simulate a disease outbreak and determine the index case.
3. Explain the importance in identifying the index case in a disease outbreak.

Prelab Questions

1. Is this lab a model of direct transmission or indirect transmission of a disease agent?
2. Why are we using the organism *Serratia marcescens* in this exercise?
3. What is the definition of *index case?*

Materials

Per student

Disposable gloves (1 pair)

Petri dish or weigh boat with jelly bean (or other type of hard candy)

Nutrient agar plate, 1

Culture

24-hour 37°C *Serratia marcescens* in Trypticase Soy broth (red pigment producing strain)

PROCEDURE

First Session

1. Each student will choose one weigh boat that contains a piece of candy covered in broth. Note the ID number of your weigh boat and write it here. _____
2. Place gloves on both hands and label your nutrient agar plate with your name and ID

number. *Alternatively, you may use one glove and place this on your nonwriting hand so you may keep record of those you contact in the exercise.*

3. When instructed to start the exercise, pick up the candy with one of your gloved hands and roll the candy around the glove to spread the broth on your palms and fingers. (Pick up the candy with your nonwriting hand so you may keep a record of those you contact in the exercise.) Be careful not to drip any of the broth on your lab benches or clothes as it may be contaminated with *Serratia marcescens*.

4. Shake hands with four students, keeping track of their ID numbers and the order in which you contact them.

5. Place your contaminated gloved hand on the surface of the NA plate. Be sure to rub the palm and fingers of the glove on the surface of the plate. Try not to break the surface of the agar.

6. Remove your gloves by turning them inside out as you pull them off of your hands. Dispose of them in the BIOHAZARD waste containers along with your candy and weigh boat.

7. Incubate your nutrient agar plates at 30°C or room temperature overnight. Most strains of *Serratia marcescens* will only produce a red pigment when grown at room temperature.

Second Session

1. Observe plates for the presence of red-pigmented colonies. *Serratia marcescens* is a Gram-negative rod that produces a prodigiosin, a type of red pigment that can range from orange to brick red in color. If this organism is present on your plate, you were infected with the "disease" agent.

2. Record the data in table 30.1 of all individuals who have red-pigmented colonies on their plates. Be sure to include the ID number of the positive plates as well as the ID numbers of all the individuals they shook hands with.

3. Based on the data, try to identify the index case. It may only be possible to narrow down the likely "suspects" to two individuals, the index case and the first person students shook hands with because this initial handshake will have a high concentration of *Serratia*. To narrow your possible choices for the index case, you may make the assumption that a likely "suspect" will have all four contacts contaminated or will be positive for red-pigmented growth on his or her plates.

EXERCISE

30

Laboratory Report: Epidemiology

Table 30.1   Epidemiology Data Results

Individuals with Red-Pigmented Colonies on Plate	Contact #1	Contact #2	Contact #3	Contact #4

Questions

1. In your class, who was the index case? Briefly explain how you determined the index case.

2. In this lab, we are assuming that everyone who comes into contact with the disease agent is infected. Is this a model of what happens in "real life"? Explain your answer.

3. On your agar plates, there may be other bacteria growing on them that are not pigmented. What is the likely source of the non-pigmented bacteria?

INTRODUCTION to Biotechnology

Watson and Crick first proposed the structure of DNA in the early 1950s. In less than 50 years, the field had mushroomed: Geneticists had isolated DNA, transferred specific genes to another organism, and determined the sequence of bases in the DNA of specific genes as well as entire bacterial genomes. These sequences have been used to identify or classify an organism and to determine evolutionary relationships.

The techniques for manipulating DNA have many applications. A particular gene may be cut from the DNA of one organism using restriction enzymes and inserted into another organism, so that the action of the gene can be studied independent of the organism. DNA manipulation is also a method for obtaining a gene product, usually a protein, in large quantities. For instance, the human gene for insulin can be cloned into yeast or bacterial DNA. These microorganisms then can be grown in huge quantities, and the insulin can be purified for use in the treatment of diabetes.

In these next four exercises, you will apply some of the techniques used in biotechnology. First you will use restriction enzymes and the polymerase chain reaction to produce patterns of DNA unique to each organism, which will be visualized on an agarose gel. In the third exercise, you will use a computer database to identify an organism from its DNA sequences of the 16S rRNA gene. In the fourth exercise, you will determine gene expression on a DNA microarray, also known as a DNA chip.

Notes:

EXERCISE

31

Identifying DNA with Restriction Enzymes and the Polymerase Chain Reaction (PCR)

Getting Started

DNA from one organism can be distinguished from the DNA of another organism using restriction enzymes (REs) and/or the polymerase chain reaction (PCR). The location of the specific sequences varies from species to species, and only identical strands of DNA are cut or amplified into the same number and size of fragments. This is the basis for comparing DNA from different organisms, one that has many applications. For example, in forensic (legal) investigations, the DNA from bloodstains can be compared with the DNA of a suspected murderer. In epidemiology investigations, the DNA from a serious *E. coli* outbreak of diarrhea can be compared to the DNA isolated from other *E. coli* strains found in food to determine the source of infection. Diseases like lyme disease and whooping cough are now diagnosed by PCR.

Enzymes are found in bacteria, which use them to degrade foreign DNA that might enter their cell. Each cell methylates its DNA by adding a methyl group at a particular site and thereby prevents its own DNA from being degraded by its own restriction enzymes. Foreign DNA entering the cell does not have this specific pattern of methylation, and so the cell cleaves it, restricting its expression in the cell. Therefore, these enzymes are called **restriction enzymes** (figure 31.1). They can be isolated from bacteria and used in the laboratory for studying and manipulating DNA. In the laboratory, DNA samples to be compared are mixed with a restriction enzyme and incubated until the enzymes have cleaved the DNA at the recognition site unique to the enzyme. The restriction enzymes' names are derived from the first letter of the bacteria's genus and the first two letters of their species. For example, *Eco*RI is the first restriction enzyme from *Escherichia coli* strain R. Not only do restriction enzymes cut DNA at very specific **sequences,** but there are many different enzymes, each with its own sequence. For example, the enzyme *Hha*I cuts at the sequence GCGC. In the two

sequences of DNA shown below, each strand is cut in two pieces because each contains GCGC; the pieces, however, differ in size between the two sequences.

Smaller (9 bp) DNA fragment

ATGGCTCAA GCGC TCACGGTAACTGCTGCATCCCGTTATACGAGCTACTT
ACCGAGTT CGCG AGT-GCCATTGACGACGTAGGGCAATATGCTCGATGA

Larger (36 bp) DNA fragment

Similarly sized DNA fragments (24 bp versus 21 bp)

ATATCGTTGAACTCCGTGTAGACT GCGC ACGTGTTACAATCCACCAAGT
TATAGCAACTTGAGGCACATCTGA CGCG TGCACAATGTTAGGTGGTTCA

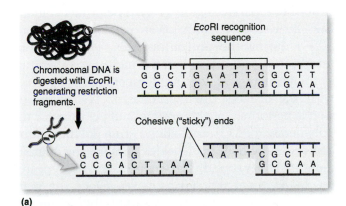

(a)

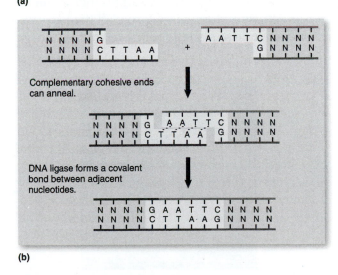

(b)

Figure 31.1 (*a*) How a restriction enzyme recognizes a sequence and cuts it. (*b*) Fragments can then anneal.

The polymerase chain reaction (PCR) **amplifies,** or increases, the quantity of DNA already present. It is performed in a programmable machine called a thermocycler (figure 31.2), which contains a metal heat block that can change temperatures at a given time. The amplification of a specific sequence or DNA area (also known as the target DNA) is achieved by designing a **primer** (also known as an oligonucleotide, or oligo) using computer software programs and having a company make the primer. The primer will only bind to the complementary DNA sequence and amplify that area (figure 31.3). Primers vary in length from 7 to 30 bases, with shorter sequences being nonspecific, whereas longer primers are generally directed to a known DNA sequence of a specific gene.

The process of PCR heats the DNA to 95°C, which is called denaturation. This causes the DNA strands to separate (figure 31.4). The primers bind to the DNA sequence at a lower temperature (around 50°C) during the annealing step. The annealing temperature will depend on the melting temperature of the primers. The melting temperature is provided with the primers when purchased. The extension step is performed at the optimal temperature for the polymerase (72°C) and allows **_Taq_ polymerase** to add the individual nucleotides to make the complementary DNA strand. These three steps are repeated for 25–45 times, known as cycles. Once the main steps have been completed, an additional 10-minute step at 72°C is performed to ensure all the bases were extended, and then the DNA is held at 4°C or put

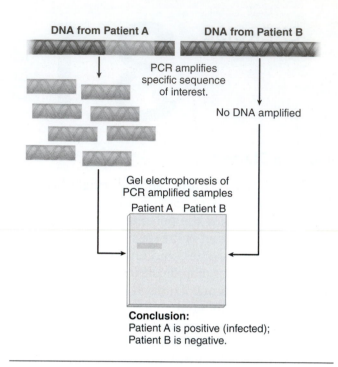

DNA from Patient A DNA from Patient B

PCR amplifies specific sequence of interest.

No DNA amplified

Gel electrophoresis of PCR amplified samples
Patient A   Patient B

Conclusion:
Patient A is positive (infected);
Patient B is negative.

Figure 31.3 How a DNA target sequence is amplified. DNA from two different patients is obtained. The primer targeting a specific disease is used and only one patient is positive for the disease, which is visualized by running the samples on an agarose gel.

into the freezer until a gel can be run. The results are then visualized by **electrophoresing** the DNA in an agarose gel (exercise 32).

The ingredients in a typical PCR reaction include *Taq* **DNA polymerase,** *Taq* DNA **buffer,** magnesium chloride (MgCl), water, **nucleotides** (dNTPs-deoxynucleoside triphosphates), DNA, and primers (oligonucleotides). The water serves as a diluent for the reaction, the buffer optimizes the pH of the solution, and MgCl provides the ions for the reaction to work properly. The primer directs the location of amplification. The *Taq* polymerase is obtained from the hot springs bacterium *Thermus aquaticus* and recognizes the dNTPs—Adenine (A), Thymine (T), Cytosine (C) and Guanine (G)—for synthesizing the complementary DNA strand.

In this exercise, you will compare the DNA of three bacterial viruses using both restriction enzymes and PCR: phage lambda, phage φ X 174, and virus X (which is either lambda or φ X 174). The restriction enzyme *Dra*I is used to cut the DNA. It recognizes the base sequences.

Figure 31.2 A thermocycler used to perform PCR, Courtesy of Anna Oller, University of Central Missouri.

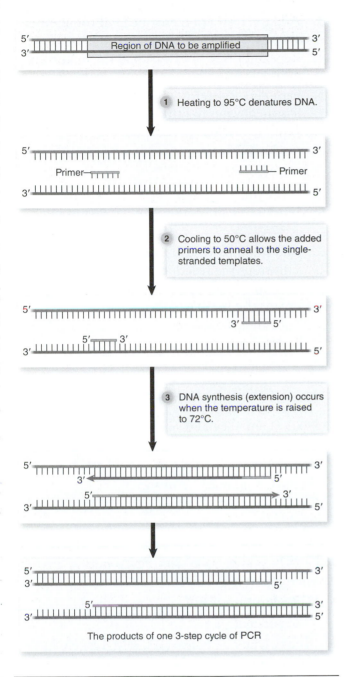

Figure 31.4 The 3 major steps involved in PCR: denaturation, annealing, and extension.

TTTAAA

AAATTT

Lambda contains thirteen sites that can be cut by this enzyme, but φ X 174 has only two. You will amplify DNA using a short primer called a randomly amplified polymorphic DNA primer. Therefore, you should be able to identify virus X as either φ X 174 or

lambda, depending on the number and size of the pieces separated by gel electrophoresis.

Note: Students should practice using micropipettors before doing this exercise and loading a practice gel (exercise 32) before the next lab due to time constraints.

Definitions

Amplify. In PCR, to make more DNA of a specific sequence and length.

Buffer solution. A salt solution formulated to maintain a particular pH.

Cell lysate. The liquid portion containing lysed cells.

DNA polymerase. Enzyme that synthesizes DNA; it uses one strand as a template to generate the complementary strand.

DNA sequencing. Determining the sequence (order) of nucleotide bases in a strand of DNA.

Electrophoresis. Technique that uses an electrical current to separate either DNA or RNA fragments or proteins.

Invert. To gently mix a small tube. Place a thumb on the bottom of the tube and middle finger on the top of tube and gently tip upside down.

Nucleotides. The deoxyribonucleic acids (adenine, thymine, guanine, and cytosine) that are used to generate a complementary DNA strand sequence.

Primer. A short (10–30 bp) DNA sequence that determines which segment of DNA will be amplified.

Restriction enzymes. Enzymes isolated from bacteria that are used to cut DNA at specific sequences.

Supernatant. The liquid portion above the pellet.

***Taq* polymerase.** Heat-stable DNA polymerase of the thermophilic bacterium *Thermus aquaticus*.

Target DNA. In the PCR procedure, the region to be amplified.

Objectives

1. Explain the use of restriction enzymes to cut DNA into segments.
2. Explain the steps in PCR and how the ingredients allow the steps to work.
3. Explain the use of DNA fingerprinting to identify DNA.

1. What will you use to transfer the liquids into the centrifuge tubes?

2. What are the 3 stages in PCR?

3. Name the main ingredients in PCR.

Materials (Restriction Enzyme Digestion)

Lambda DNA

φ X 174 DNA

Unknown phage (either lambda or φ X 174)

Restriction enzyme *Dra*I (or another enzyme)

Sterile 0.5 ml microcentrifuge tubes or Eppendorf tubes

Micropipettors

TE buffer

Stop mix (includes tracking dye)

Water bath or incubator

Microfuge/Centrifuge

Disposable gloves

Distilled water (sterile)

Ice

Restriction Enzyme Digestion

PROCEDURE

You may use the phage DNA or your instructor may have you isolate your own DNA from a bacterium to use in the restriction enzyme digestion and PCR. The instructions for isolating DNA are found at the end of this lab exercise.

Restriction Enzyme Digestion

1. Obtain 3 sterile tubes and label with a marker. Add _____ μl sterile distilled water (as determined by the instructor) to a sterile Eppendorf tube on ice, and then add 1 μl 10X TE buffer. Buffer is added before the enzyme so that the conditions are immediately optimal for the enzyme.

2. Add _____ μl lambda DNA to the tube. The amount added is usually 0.5–1.0 μg/μl. The instructor will indicate how much you should add.

3. Add the restriction enzyme *Dra*I (or another enzyme).

4. Repeat for φ X 174 DNA and the unknown virus DNA.

5. Each tube contains:
   Distilled water to bring total volume to 10 μl, including enzyme.

   10X TE buffer

   DNA

   Enzyme

6. Mix and incubate the tubes for 30 minutes at 37°C.

7. Add stop mix (usually 1 μl of 10X, or EDTA). This stops the reaction. Some stop mixes also contain a loading dye that helps the sample sink to the bottom of the well.

8. Tubes may be stored at −20°C until a gel can be run.

Materials (Polymerase Chain Reaction)

DNA (Lambda and φ X 174 or DNA you isolated)

Primer 5'-CCGCAGCCA-3'(100 mM)

Sterile 0.5 ml microcentrifuge tubes

0.5 ml or 0.2 ml microcentrifuge PCR tubes, sterile

Thermocycler (PCR tubes need to fit the heat block)

Micropipettors

PCR Mastermix OR:
  Taq polymerase (5 U μl^{-1})
  Taq polymerase buffer (100 mM Tris-HCL, 500 mM KCl, 0.8% Nonidet P-40)
  dNTPs (10 mM)
  MgCl (25 mM)

Sterile double distilled purified water (PCR reagent grade)

Ice bath

Microfuge/Centrifuge

Permanent fine point marker for writing on plastic tubes (Sharpies will melt off)

Disposable gloves

Polymerase Chain Reaction

Always pipet slowly so you do not create bubbles. ALL TUBES MUST be kept on ice!

1. Label your tube on the top and side with a marker. Wear gloves while you pipet to prevent your cells from contaminating the tubes. Keep the tube lid shut to prevent contaminating your tubes from the air.

2. You will add the following items into a 0.5 microcentrifuge tube in the following order. The list on the left is the predetermined reaction, but your instructor may modify it. If so, write the amounts in the list on the right. If your instructor has aliquotted the primer and *Taq* polymerase into tubes, gently pipet 5 µl of the water solution into the tube, add 1 µl to the pipet dial, and then pipet the solution back into the water.

17 µl Water	_____ µl Water
0.25 µl Primer	_____ µl Primer
6 µl PCR Mastermix OR:	_____ µl Mastermix OR
2.5 µl *Taq* DNA BUFFER	_____ µl *Taq* DNA BUFFER
2 µl MgCl	_____ µl MgCl
1 µl dNTPs	_____ µl dNTPs
0.15 µl *Taq* DNA polymerase	_____ µl *Taq* DNA polymerase
2 µl DNA	_____ µl DNA

   25 µl total reaction

3. Now pipet the 25 µl into the 0.2 (or 0.5 ml) thin-walled PCR tube. After adding all ingredients together, make sure to balance the centrifuge and pulse spin the tube for 3–5 seconds.

4. Add tubes to thermocycler (already programmed).

5. The thermocycler was programmed for 30 cycles according to the following or your own formula:

   95°C for 1 minute (denature)   95°C _____

   50°C for 2 minutes (anneal)   __°C _____

   72°C for 1 minute (extend)   72°C

6. The following were then programmed in for 1 cycle:

   72°C for 10 minutes (extension)   72°C for 10 minutes

   4°C for indefinitely   4°C indefinitely

7. The PCR run will take about 3 hours to complete. The samples can then be stored at −20°C.

DNA Isolation

The DNA of Gram-negative bacteria can be isolated using commercial kits that you buy, or you can use a standard protocol like the one listed below.

Materials (DNA Isolation)

Microcentrifuge tubes (1.5 ml)

Gloves

Micropipettors

Gram-negative DNA isolation kit OR

  Trypticase soy or nutrient broth (depending on the bacterium) containing log phase bacteria like *Escherichia coli*, *Citrobacter*, *Klebsiella*, etc.

1 mg/ml of lysozyme dissolved in filter sterilized STET buffer (5% (v/v)) Triton X-100

50 mM Tris-HCl, pH 8.0

50 mM EDTA, pH 8.0, 8% (w/v) sucrose)

TE buffer (10 mM Tris-HCl pH 8, 1 mM EDTA)

70% Ethanol

Isopropanol

Water bath (boiling)

Microcentrifuge rack that can be placed into the water or a floatie

Sterile wooden toothpicks

Permanent fine-point marker for writing on plastic tubes

Glass beaker, container with bleach, or autoclave bag

Vortexer (optional)

Forceps

Ice

DNA Isolation

PROCEDURE

Because we are isolating DNA in an open room, keep microcentrifuge tubes closed as much as possible so you do not have contamination. Further, try not to cough, sneeze, breathe, etc. directly above/into the tubes.

Tip: Always centrifuge your tubes in the same direction, with the hinge facing the back of the rotor (away) and the lid facing the middle of the rotor. This way, any precipitate will be seen directly under the hinge so you know exactly where to look for your DNA. Do not worry if you do not see the DNA. If you see a very large pellet, it most likely is not DNA and is contaminating proteins.

1. The day before the isolation, inoculate the bacterium into a trypticase soy or nutrient broth tube and grow for 12–24 hours at 37°C.

2. Pipet the bacteria from the broth tube to a 1.5 ml microcentrifuge tube, label the tube with a marker, and centrifuge at 12,000 rpm for 1 minute to pellet the cells.

   a. If a small pellet is seen, pipet the **supernatant** off and dispense into a glass beaker or other item for disposal. Pipet another 1 ml of the broth into the 1.5 ml tube and centrifuge again.

3. Pipet off the supernatant and discard into a glass beaker.

4. Pipet 20 μl of lysozyme to the pellet, and gently pipet up and down to break up the pellet.

5. Make sure the microcentrifuge cap is firmly locked down into place.

6. Place the centrifuge tube containing the lysozyme with STET into a rack or floatie and place into the boiling water for 1 minute. Make sure the tube is at least halfway submerged in the boiling water.

7. Using forceps, remove the microcentrifuge tube from the boiling water and place on ice for 2 minutes. This contains the **cell lysate.**

8. Centrifuge the tubes for 10 minutes at 12,000 rpm. You should see a large pellet. The pellet contains the cell components like proteins, membranes, etc.

9. Carefully pipet off the supernatant into a new, labeled microcentrifuge tube. The supernatant contains the DNA.

10. Pipet 200 μl of refrigerated isopropanol into the microcentrifuge tube containing the supernatant.

11. **Invert** the tube 50 times.

12. Centrifuge the tubes for 5 minutes at 12,000 rpm. You may be able to see a small pellet of DNA.

13. Pipet off the supernatant into the glass beaker.

14. Add 300 μl of refrigerated 70% ethanol to the tube. Invert the tube 2 times.

15. Centrifuge the tubes for 1 minute at 12,000 rpm.

16. Open the lid to the centrifuge tube. **Quickly and in 1 motion,** invert the tube onto a piece of towel. Gently tap to remove the excess ethanol.

17. Allow the tubes to dry for about 15 minutes. If you could see a DNA pellet before, it will now become clear as it dries.

18. Resuspend the DNA in 25–50 μl of TE buffer by pipetting up and down. Allow to set at room temperature overnight or at 55°C for 4 hours. This step allows the DNA to regain its conformation.

EXERCISE

31

Laboratory Report: Identifying DNA with Restriction Enzymes and the Polymerase Chain Reaction (PCR)

Questions

1. Would you have the same number of fragments from each phage if you used a different restriction enzyme? Why or why not?

2. Name and describe the purpose of each of the main ingredients in an enzyme digestion.

3. What would happen if the PCR primer was not designed properly?

4. Name and explain two things that might inhibit a PCR reaction from working properly.

32

Electrophoresis

Getting Started

Electrophoresis is a widely used molecular biology technique based upon the principle of molecular attraction of negative and positively charged molecules (similar to North–South magnets). Electrophoresis uses an electrical current to separate out molecules based on their base pair size. It is used to visualize results of procedures, such as after performing PCR. One end of the electrophoresis apparatus is negatively charged and the electrodes appear black, whereas the other end is positively charged and will have red electrodes (figure 32.1). DNA is negatively charged, so it will be attracted to the positive electrical charge. This is why we add DNA to the negative end of the gel. If the gel runs too long, the samples run off the bottom of the gel and diffuse into the buffer solution, which cannot be recovered.

Different mediums can be used in electrophoresis such as acrylamide for separating proteins, and agarose for separating DNA or RNA. **Agarose** is a purified agar that will solidify once it has been boiled and allowed to cool. It is porous to allow molecules to move through it, and different concentrations of agar can be used to make the gel more or less porous. Small molecules move through the gel faster than larger molecules. The faster a molecule travels, the farther away from the loading wells it will move; thus, they are usually toward the bottom or end of the gel. A thicker, less porous gel is used to separate large molecules (thousands of base pairs) from one another, whereas a thinner, more porous gel is used to separate smaller-sized molecules (hundreds of base pairs) from one another. Once the agarose solution has boiled, it will need to cool for 5–10 minutes so the heat does not melt or warp the plastic trays. The plastic trays serve as a mold, similar to a mold you would use to make Jello™. The agarose solution is then poured into the trays and **combs** are added. Once the gel has solidified, the comb is pulled out, leaving a well. The DNA is then added to the wells. Combs

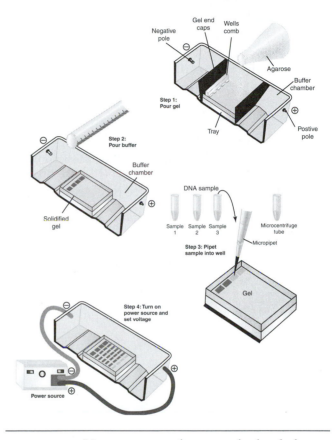

Figure 32.1 How agarose gels are made, loaded, and run. Step 1: The agarose is melted in buffer by heating, and poured into the mold containing a comb that creates the wells when it is removed upon gel cooling. The wells are made on the end of the gel that will be towards the negatively charged end. Step 2: Once the gel has cooled, the comb is pulled out, and buffer is poured over the gel. Step 3: The samples are loaded into the wells of the gel, which is still in the gel box. Step 4: The electricity is turned on to allow the molecules to separate based on charge and size.

may have varying numbers of projections, such as between 5 and 10. The combs may also have different thicknesses. Thicker combs allow more DNA to be added, so visualization of small quantities of

DNA is enhanced. Thinner, longer combs allow for neater bands so if two bands appear that are similar in size, they are easier to separate and visualize.

Different types of **buffers** can be used, depending upon the type and size of molecules being separated. A 1× TAE (tris-acetate-EDTA) solution will be used for DNA. For RNA separation, a TBE (tris-borate-EDTA) solution could be used instead. Agarose is added to the 1× TAE buffer and is boiled. The same TAE buffer is also the solution used to run the gel in the box so the molecules and charge between the gel and the solution will be consistent. DNA can be viewed in the gel a few different ways. The DNA can be stained with a dye (methylene blue or a fluorescent dye) in a glass or plastic tub after the gel has been run and then de-stained in distilled water. Second, a solution of ethidium bromide or a fluorescent dye can be added directly to the flask of agarose before it is poured, or it can be directly added to the electrophoresis buffer at the positive end. Ethidium bromide is positively charged so it

migrates to the negative end, and the DNA traps it, so when the **UV transilluminator** is turned on, the DNA can be seen in the gel. The new fluorescent dyes do not have the same safety concerns as ethidium bromide and are excellent at detecting small amounts of DNA.

The DNA samples will have to be mixed with a 6× **loading dye** that contains a dye (bromophenol blue) or dyes that are dissolved in a solution that will bind to the DNA, weigh it down, and allow it to fall into the wells. If a loading dye is not used, then the DNA will float into the buffer solution and cannot be recovered. The dye provides an estimation of how far down the gel the molecules have traveled. DNA bands will be seen, which are fragments of the DNA (figure 32.2).

A standard must be added to the first well for each gel run. Different **ladders** exist so you can compare the samples to the standard to determine the number of base pairs present. We use a 1 kb (kilobase) DNA ladder (standard), which means the lowest band is

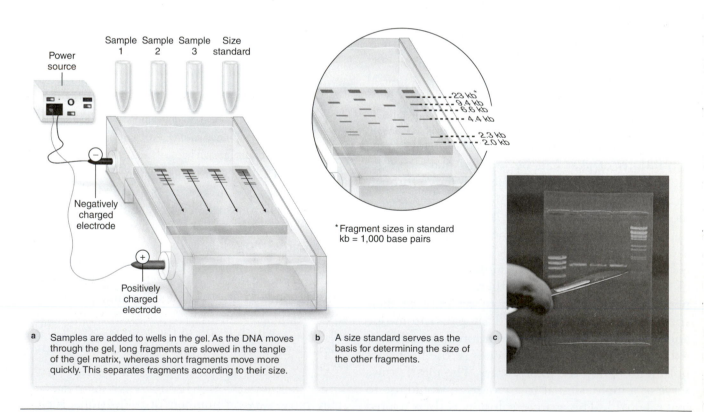

a Samples are added to wells in the gel. As the DNA moves through the gel, long fragments are slowed in the tangle of the gel matrix, whereas short fragments move more quickly. This separates fragments according to their size.

b A size standard serves as the basis for determining the size of the other fragments.

c

Figure 32.2 (a) How electrophoresis separates molecules based on size. (b) A standard is loaded onto every gel so fragment size can be determined. (c) Visualization of DNA bands on a UV transilluminator. © Richard T. Nowitz/Photo Researchers, Inc.

1,000 base pairs (1 kilobase = 1,000 base pairs). The ladder is also mixed with 6× loading dye to keep it in the well.

You will run an agarose gel using your PCR and restriction enzyme digestion samples from the previous lab and estimate the band fragment sizes.

Definitions

Agarose. Highly purified form of agar used in gel electrophoresis.

Buffer. Substance in a solution that acts to prevent changes in pH.

Comb. A piece of plastic containing a variable number of projections that forms the indentations, or wells, when the gel solidifies and the comb is removed. DNA is added to the wells.

Electrophoresis. Technique that uses an electrical current to separate either DNA or RNA fragments or proteins.

Gel electrophoresis. Technique that uses an electrical current to separate DNA or RNA fragments according to size by drawing them through a slab of gel, which has the consistency of very firm gelatin.

Ladder. A standard marker run on each gel with known base pair band sizes. Sample sizes are compared to the ladder.

Loading dye. A dye that weighs down the DNA so it stays in the wells and provides an estimation of band location.

UV transilluminator. A piece of equipment that uses a wavelength of UV light to excite the molecules in the gel so samples can be visualized.

Objectives

1. List the major steps in performing electrophoresis.
2. Explain how electrical current can be used to move molecules.
3. Interpret the size of molecules electrophoresed.

Prelab Questions

1. What size molecules will be seen at the top of the gel by the wells?

2. Why must you add a loading dye to the DNA solution?

3. What is the purpose of the DNA ladder?

Materials

Electrophoresis apparatus

Combs (10 wells)

2 large binder clips to hold combs in place

Small gel trays (holds 50 ml)

Masking tape (if trays do not have end caps)

Agarose

1× TAE (tris-acetate-EDTA) solution

Micropipettors (1–2 µl, 5 µl and 15 µl)

Hothands or potholders

DNA ladder

6× loading dye

Ethidium bromide or a fluorescent dye (ethidium bromide = 1 µl of a 10× solution per gel)

Gloves

UV transilluminator

DNA (from previous PCR and restriction enzyme digestion)

PROCEDURE

1. Prepare a 1% agarose gel. This may already be done for you.
   a. Add 0.5 g agarose to 50 ml 1×TAE buffer in a 150 ml flask. Add a Kimwipe to the top of the flask to help prevent evaporation.
   b. Boil until the agarose molecules completely disappear. Watch the flask so it does not boil over. (If microwaving, boiling occurs at about 30 seconds. Open the microwave to swirl and avoid boiling over; continue boiling until the solution no longer bubbles, about 1 minute.)

c. Remove the flask with a potholder and allow to cool for about 5 minutes on the benchtop.

d. While the flask is cooling, set up the gel tray by placing the end caps on the tray (or taping the ends with masking tape) and placing the 10-well comb at the end that will be aligned with the negative (black) electrode.

e. Put on gloves. Add 1 μl of ethidium bromide (or other dye) to the solution, and pipet up and down to remove as much of the ethidium bromide as possible. Discard in the appropriate container.

f. Pour the gel into the tray and allow to solidify. (You may pour the gels in the refrigerator to facilitate solidifying.) Do not move gels once they are poured until they have turned opaque white, signaling they are solidified.

g. Remove the binder clips and then the comb by pulling straight up.

h. Remove the end caps or masking tape. You will need to keep your gloved fingers on the ends of the gel or it may fall out of the tray.

2. Place the gel into the electrophoresis apparatus and be sure the wells containing your sample are at the negative end.

3. Add enough 1× TAE buffer to the electrophoresis apparatus to cover the wells (look in from the side to ensure full coverage).

4. Each group can add 2 μl of 6× loading dye to the 25 μl PCR sample. If you have a 1 μl pipet, you will have to do the transfer twice. Pipet up and down to mix the sample.

5. Each group can add 2 μl of 6× loading dye to the enzyme digestion reaction. If you have a 1 μl pipet, you will have to do the transfer twice. Pipet up and down to mix the sample.

Note: One set of gels will be used for the PCR samples, and the other gels will be used for the restriction enzyme digestion. Write your sample name on the board next to the lane you used.

6. Add 5 μl of a 1 kb DNA ladder to lane 1 of each gel by aiming the pipet tip over the well and **slowly** pushing the pipet plunger to dispense the solution into the gel. (If you push the plunger too quickly, your sample may be pushed out of the well and into the buffer solution.) Do not release the plunger until you have removed the pipet tip from the solution or it will suck the buffer and some of the sample back up.

7. Add 15 μl of the PCR sample to lane 2, etc., of the appropriate gel. Each group should continue until all groups have added their samples.

8. Add 15 μl of the enzyme digestion reaction to lane 2, etc., of the appropriate gel. Each group should continue until all groups have added their samples.

9. Plug the electrical cords into the gel and into the electrical box.

10. Allow your gel to run for about an hour at 80 volts.

11. Once the gel is phoresed, turn off the electrical current and disconnect the electrical cords.

12. Wearing gloves, carefully remove the gel tray and move the gel onto the UV transilluminator. Wearing goggles, turn off the lights, and turn on the UV transilluminator to visualize the DNA bands. Depending on the dye used, prolonged exposure to the UV light may cause the bands to fade.

13. Photograph the gel or make a drawing of the bands.

14. Depending upon the expected PCR product size, a smudged band at the very bottom of each sample may be leftover ingredients (color plate 50).

EXERCISE

32

Laboratory Report: Electrophoresis

Results and Questions

1. Record the gel results for the **PCR** sample in the table. Record the base pair size for band 1, band 2, band 3, etc.

	Ladder	Lane 1	Lane 2	Lane 3	Lane 4	Lane 5	Lane 6	Lane 7	Lane 8	Lane 9
Sample Identity Total # Bands Seen										
Band 1 size										
Band 2 size										
Band 3 size										
Band 4 size										
Band 5 size										

2. Record the gel results for the **restriction enzyme digestion** sample in the table. Record the base pair size for band 1, band 2, band 3, etc.

	Ladder	Lane 1	Lane 2	Lane 3	Lane 4	Lane 5	Lane 6	Lane 7	Lane 8	Lane 9
Sample Identity Total # Bands Seen										
Band 1 size										
Band 2 size										
Band 3 size										
Band 4 size										
Band 5 size										

3. What is the probable identity of the phage X?

4. How were you able to estimate the size of the DNA fragments?

5. How bright were the bands in comparison to the ladder (lighter, the same, a little brighter/darker, much darker)?

6. Explain reasons why you may not have had any bands present, if you didn't see any.

Getting Started

Identifying and classifying bacteria have always been more difficult than identifying and classifying plants and animals. Bacteria have very few differences in their structure, and while they are metabolically extremely diverse, it has not been clear which of these characteristics are the most important for identification or grouping. For example, is nitrogen fixation more significant than anaerobic growth, or endospore formation more important than photosynthesis?

The ability to sequence the DNA of microorganisms has offered a new solution. The closer organisms are related to each other, the more similar their **DNA nucleotide sequences.** Certain genes common to all bacteria can be sequenced and compared. What genes should be compared? The DNA coding for a part of the ribosome, namely the **16S** ribosome portion, is a good choice. Ribosomes are critically important for protein synthesis, and any mutation is quite likely to be harmful. Some mutations, however, are neutral or perhaps even advantageous, and these mutations then become part of the permanent genome. These sequences change very slowly over time and are described as highly conserved. The 16S rRNA DNA segment is found in all organisms (slightly larger in eukaryotes) with the same function, so the sequences can easily be compared.

This approach has had exciting results. Some specific base sequences are always found in some organisms and not others. These are called **signature sequences.** Also, new relationships among bacteria can be determined by comparing sequences of organisms base by base by means of computer programs. The more the sequences diverge, the more the organisms have evolved from one another.

Perhaps even more useful, these sequences can be used to identify bacteria. The sequences of at least 16,000 organisms are in public databases, with new sequences continually added. If a new, unknown sequence is submitted to a database management computer, it will respond in seconds with the most likely identification of the species containing the sequence. As the number of rRNA sequences grows in the database, so do the possible applications of this information. For eample, in a clinical setting, rRNA sequence data can now be used to identify a pathogen more quickly than by the use of standard culture techniques. Faster identification of an organism enables clinicians to treat disease-causing organisms more efficiently, often with a better outcome for the patient.

This exercise will give you a chance to send the DNA sequence of the 16S ribosomal RNA of an organism to the Ribosomal Database Project (RDP) and to identify the organism. Because many schools do not have the resources for determining the specific sequences, you will be given the sequences, which you can enter on the Web and immediately receive an identification. You may also look for other sequences posted on the Web or in microbiological journals, such as *Journal of Bacteriology*, published by the American Society for Microbiology.

Definitions

DNA nucleotide sequence. The order that bases are found in a piece of DNA.

16S. *S* is an abbreviation for Svedberg. It is a unit of mass that is measured by the rate a particle sediments in a centrifuge. The prokaryotic ribosome is made up of two main parts: 30S and 50S. The 30S particle is made up of the 16S rRNA plus twenty-one polypeptide chains.

Signature sequences. DNA sequences of about 5–10 bases long found at a particular location in the 16S rRNA. These DNA sequences are unique to Archaea, Bacteria, or Eukarya (eukaryotes).

Objectives

1. Explain the importance of the 16S rRNA sequence for the identification of organisms.

2. Interpret rRNA data using the Ribosomal Database Project.

Prelab Questions

1. What sequence information is used to identify organisms in this lab?

2. What are signature sequences?

3. If an organism is related to another, will the rRNA sequence be more similar or different?

PROCEDURE

1. Open your Internet browser (Netscape or Explorer, for example) on the computer.

2. Type in the URL http://rdp.cme.msu.edu

3. Under "RDP Analysis Tools," click "Sequence Match."

4. Type the sequence in the box labeled "Cut and Paste Sequence(s)." This is easier to do if one person reads the sequence while the other person types in the letters. Sometimes only a few hundred bases are necessary for an identification.

5. Click "Submit."

6. Read identification under "Hierarchy View."

Hint: The organisms were studied in

1. Exercise 22

2. Exercise 22

3. Exercise 23

4. Exercise 13 (genus only) Photosynthetic bacteria are not found in any exercise.

Organism 1

```
   1    tctctgatgt tagcggcgga cgggtgagta acacgtggat aacctaccta taagactggg
  61    ataacttcgg gaaaccggag ctaataccgg ataatatttt gaaccgcatg gttcaaaagt
 121    gaaagacggt cttgctgtca cttatagatg gatccgcgct gcattagcta gttggtaagg
 181    taacggctta ccaaggcaac gatgcatagc cgacctgaga gggtgatcgg ccacactgga
 241    actgagacac ggtccagact cctacgggag gcagcagtag ggaatcttcc gcaatgggcg
 301    aaagcctgac ggagcaacgc cgcgtgagtg atgaaggtct tcggatcgta aaactctgtt
 361    attagggaag aacatatgtg taagtaactg tgcacatctt gacggtacct aatcagaaag
 421    ccacggctaa ctacgtgcca gcagccgcgg taatacgtag gtggcaagcg ttatccggaa
 481    ttattgggcg taaagcgcgc gtaggcggtt ttttaagtct gatgtgaaag cccacggctc
 541    aaccgtggag ggtcattgga aactggaaaa cttgagtgca gaagaggaaa gtggaattcc
 601    atgtgtagcg gttaaatgcg cagagatatg gaggaacacc agtggcgaag gcgactttct
 661    ggtctgtaac tgacgctgat gtgcgaaagc gtgggaatca aacaggatta gataccctgg
 721    tagtccacgc cgtaaacgat gagtgctaag tgttaggggg tttccgcccc ttagtgctgc
 781    agctaacgca ttaagcactc cgcctgggga gtacgaccgc aaggttgaaa ctcaaaggaa
 841    ttgacgggga cccgcacaag cggtggagca tgtggtttaa ttcgaagcaa cgcgaagaac
 901    cttaccaaat cttgacatcc tttgacaact ctagagatag agccttcccc ttcgggggac
 961    aaagtgacag gtggtgcatg gttgtcgtca gctcgtgtcg tgagatgttg ggttaagtcc
1021    cgcaacgagc gcaaccctta agcttagttg ccatcattaa gttgggcact ctaagttgac
1081    tgccggtgac aaaccggagg aaggtgggga tgacgtcaaa tcatcatgcc ccttatgatt
1141    tgggctacac acgtgctaca atggacaata caaagggcag cgaaaccgcg aggtcaagca
1201    aatcccataa agttgttctc agttcggatt gtagtctgca actcgactac atgaagctgg
1261    aatcgctagt aatcgtagat cagcatgcta cggtgaatac gttcccgggt cttgtacaca
1321    ccgcccgtca caccacgaga gtttgtaaca
```

Organism 2

```
1     taacacgtgg ataacctacc tataagactg ggataacttc gggaaaccgg agctaatacc
61    ggataatata ttgaaccgca tggttcaata gtgaaagacg gttttgctgt cacttataga
121   tggatccgcg ccgcattagc tagttggtaa ggtaacggct taccaaggca acgatgcgta
181   gccgacctga gagggtgatc ggccacactg gaactgagac acggtccaga ctcctacggg
241   aggcagcagt agggaatctt ccgcaatggg cgaaagcctg acggagcaac gccgcgtgag
301   tgatgaaggt cttcggatcg taaaactctg ttattaggga agaacaaatg tgtaagtaac
361   tatgcacgtc ttgacggtac ctaatcagaa agccacggct aactacgtgc
```

Organism 3

```
1     gcctaataca tgcaagtaga acgctgagaa ctggtgcttg caccggttca aggagttgcg
61    aacgggtgag taacgcgtag gtaacctacc tcatagcggg ggataactat tggaaacgat
121   agctaatacc gcataagaga gactaacgca tgttagtaat ttaaaagggg caattgctcc
181   actatgagat ggacctgcgt tgtattagct agttggtgag gtaaaggctc accaaggcga
241   cgatacatag ccgacctgag agggtgatcg gccacactgg gactgagaca cggcccagac
301   tcctacggga ggcagcagta gggaatcttc ggcaatgggg gcaaccctga ccgagcaacg
361   ccgcgtgagt gaagaaggtt ttcggatcgt aaagctctgt tgttagagaa gaatgatggt
421   gggagtggaa aatccaccaa gtgacggtaa ctaaccagaa agggacggct aactacgtgc
481   cagcagccgc ggtaatacgt aggtcccgag cgttgtccgg atttattggg cgtaaagcga
541   gcgcaggcgg tttttaagt ctgaagttaa aggcattggc tcaaccaatg tacgctttgg
601   aaactggaga acttgagtgc agaaggggag agtggaattc catgtgtagc ggtgaaatgc
661   gtagatatat ggaggaacac cggtggcgaa agcggctctc tggtctgtaa ctgacgctga
721   ggctcgaaag cgtgggggagc aaagaggatt agataccctg gtagtccacg ccgtaaacga
781   tgagtgctag gtgttaggcc ctttccgggg cttagtgccg gagctaacgc attaagcact
841   ccgcctgggg agtacgaccg caaggttgaa actcaaagga attgacgggg gcccgcacaa
901   gcggtggagc atgtggttta attcgaagca acgcgaagaa ccttaccagg tcttgacatc
961   ccgatgcccg ctctagagat agagttttac ttcggtacat cggtgacagg tggtgcatgg
1021  ttgtcgtcag ctcgtgtcgt gagatgttgg gttaagtccc gcaacgagcg caaccccctat
1081  tgttagttgc catcattaag ttgggcactc tag
```

Organism 4

```
1     ggtaccactc ggcccgaccg aacgcactcg cgcggatgac cggccgacct ccgcctacgc
61    aatacgctgt ggcgtgtgtc cctggtgtgg gccgccatca cgaagcgctg ctggttcgac
121   ggtgttttat gtaccccacc actcggatga gatgcgaacg acgtgaggtg gctcggtgca
181   cccgacgcca ctgattgacg ccccctcgtc ccgttcggac ggaacccgac tgggttcagt
241   ccgatgccct taagtacaac agggtacttc ggtggaatgc gaacgacaat ggggccgccc
301   ggttacacgg gtggccgacg catgactccg ctgatcggtt cggcgttcgg ccgaactcga
361   ttcgatgccc ttaagtaata acgggtgttc cgatgagatg cgaacgacaa tgaggctatc
421   cggcttcgtc cgggtggctg atgcatctct tcgacgctct ccatggtgtc ggtctcactc
481   tcagtgagtg tgattcgatg cccttaagta ataacgggcg ttacgaggaa ttgcgaacga
541   caatgtggct acctggttct cccaggtggt taacgcgtgt tcctcgccgc cctggtgggc
601   aaacgtcacg ctcgattcga gcgtgattcg atgcccttaa gtaataacgg ggcgttcggg
661   gaaatgcgaa cgtcgtcttg gactgatcgg agtccgatgg gtttatgacc tgtcgaactc
721   tacggtctgg tccgaaggaa tgaggattcc acacctgcgg tccgccgtaa agatggaatc
781   tgatgttagc cttgatggtt tggtgacatc caactggcca cgacgatacg tcgtgtgcta
841   agggacacat tacgtgtccc cgccaaacca agacttgata gtcttggtcg ctgggaacca
901   tcccagcaaa ttccggttga tcctgccgga ggccattgc
```

Organism 5

```
1     agagtttgat cctggctcag agcgaacgct ggcggcaggc ttaacacatg caagtcgaac
61    gggcgtagca atacgtcagt ggcagacggg tgagtaacgc gtgggaacgt accttttggt
121   tcggaacaac acagggaaac ttgtgctaat accggataag cccttacggg gaaagattta
181   tcgccgaaag atcggcccgc gtctgattag ctagttggtg aggtaatggc tcaccaaggc
241   gacgatcagt agctggtctg agaggatgat cagccacatt gggactgaga cacggcccaa
301   actcctacgg gaggcagcag tggggaatat tggacaatgg gcgaaagcct gatccagcca
361   tgccgcgtga gtgatgaagg ccctagggtt gtaaagctct tttgtgcggg aagataatga
421   cggtaccgca agaataagcc ccggctaact tcgtgccagc agccgcggta atacgaaggg
481   ggctagcgtt gctcggaatc actgggcgta aagggtgcgt aggcgggttt ctaagtcaga
541   ggtgaaagcc tggagctcaa ctccagaact gcctttgata ctggaagtct tgagtatggc
601   agaggtgagt ggaactgcga gtgtagaggt gaaattcgta gatattcgca agaacaccag
661   tggcgaaggc ggctcactgg gccattactg acgctgaggc acgaaagcgt ggggagcaaa
721   caggattaga taccctggta gtccacgccg taaacgatga atgccagccg ttagtgggtt
781   tactcactag tggcgcagct aacgctttaa gcattccgcc tggggagtac ggtcgcaaga
841   ttaaaactca aaggaattga cggggggcccg cacaagcggt ggagcatgtg gtttaattcg
901   acgcaacgcg cagaacctta ccagcccttg acatgtccag gaccggtcgc agagacgtga
961   ccttctcttc ggagcctgga gcacaggtgc tgcatggctg tcgtcagctc gtgtcgtgag
1021  atgttgggtt aagtcccgca acgagcgcaa ccccgtcct tagttgctac catttagttg
1081  agcactctaa ggagactgcc ggtgataagc cgcgaggaag gtggggatga cgtcaagtcc
1141  tcatggccct tacgggctgg gctacacacg tgctacaatg gcggtgacaa tgggaagcta
1201  aggggtgacc cttcgcaaat ctcaaaaagc cgtctcagtt cggattgggc tctgcaactc
1261  gagcccatga gttggaatc gctagtaatc gtggatcagc atgccacggt gaatacgttc
1321  ccgggccttg tacacaccgc ccgtcacacc atgggagttg gctttacctg aagacggtgc
1381  gctaaccagc aatggggggca gccggccacg gtagggtcag cgactggggt gaagtcgtaa
1441  caaggtagcc gtaggggaac ctgcggctgg atcacctcct t
```

33–4 Exercise 33 Identification of Bacteria Using the Ribosomal Database Project

EXERCISE

33

Laboratory Report: Identification of Bacteria Using the Ribosomal Database Project

Results

Organism	Bacteria or Archaea
1 _____	_____
2 _____	_____
3 _____	_____
4 _____	_____
5 _____	_____

Questions

1. What is an advantage of identifying an organism using the Ribosomal Database Project?

2. What is a disadvantage?

3. Can you identify a clinical application based on rRNA identification of an organism?

EXERCISE 34

Microarray

Getting Started

Microarrays, which are also known as *DNA chips*, determine if genes are being expressed at a given point in time. Microarrays are advantageous over other molecular biology techniques in that thousands of genes can be scanned quickly, saving time. Microarrays are currently being used in determining gene expression levels between healthy and diseased tissues, and different growth conditions for certain bacterial species. For example, their use could compare gene expression of *Escherichia coli* grown in glucose to that being grown in lactose. It could also determine sequence differences between bacteria. Currently, microarray slides must be custom-made by a company or research lab and are fairly costly. Some scientists believe that eventually microarrays will be used to diagnose diseases such as cancer or microbial diseases and help direct their respective treatments.

The mRNA is isolated from the microbe and converted to single-stranded complementary DNA (cDNA) via reverse transcription. The cDNA is then fluorescently labeled. One set of cDNA is labeled with a green fluorescent marker, whereas the other one is labeled with a red fluorescent marker. The genome of the microbe or specific gene sequence must be known, as that is the sequence that will be placed in spots on the microarray slide. A **probe** is the nucleotide sequence that will bind, or **hybridize,** to its complementary sequence. If the cDNA sequence is complementary to the known sequence, then it will bind. The excess nucleotides that did not hybridize are washed away. The spots where the cDNA hybridized to the known sequences will fluoresce when excited by a laser (figure 34.1). The amount of hybridizing will also determine how bright the fluorescence will be. For example, if the entire DNA sequence is complementary, then it will fluoresce brighter than if only a few bases are complementary. Areas will fluoresce red or green, depending on which cDNA bound to the probe. Areas where both cDNAs hybridized will fluoresce yellow,

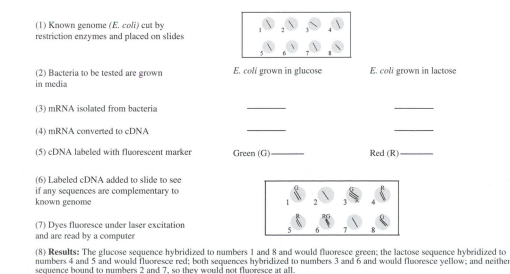

(1) Known genome (*E. coli*) cut by restriction enzymes and placed on slides

(2) Bacteria to be tested are grown in media

(3) mRNA isolated from bacteria

(4) mRNA converted to cDNA

(5) cDNA labeled with fluorescent marker

(6) Labeled cDNA added to slide to see if any sequences are complementary to known genome

(7) Dyes fluoresce under laser excitation and are read by a computer

(8) **Results:** The glucose sequence hybridized to numbers 1 and 8 and would fluoresce green; the lactose sequence hybridized to numbers 4 and 5 and would fluoresce red; both sequences hybridized to numbers 3 and 6 and would fluoresce yellow; and neither sequence bound to numbers 2 and 7, so they would not fluoresce at all.

Figure 34.1 The concept of a DNA microarray.

Courtesy of Anna Oller, University of Central Missouri.

and areas where neither bound will appear dark. A computer scanner reads the slide and determines if a gene is expressed; it will also quantitate the amount of expression.

Although microarray slides have to be read by an analyzer, the activity today will let you visualize the results. (**Note:** The colors you see today are NOT representative of the reds and greens you would actually see!) The slide you will use today will look at 6 genes. The hybridizing solution then allows for binding of complementary nucleotides.

Definitions

Hybridize. The binding of two complementary strands of DNA.
Microarray. A molecular biology technique that uses a slide containing many known nucleotide sequences to determine gene expression.
Probe. A single stranded piece of nucleic acid that has been tagged with a detectable marker.
Reverse transcription. A process that converts RNA to complementary DNA (cDNA) using the enzyme reverse transcriptase.

Objectives

1. Describe the concept of a microarray.
2. Explain how a microarray can detect gene expression.
3. Classify which genes in the microarray were expressed.

Prelab Questions

1. What is the purpose of a microarray?
2. List the main steps of a microarray from the beginning to detection.
3. If you see yellow on a computerized microarray spot, what does it mean?

Materials

Water bath or hot plate to heat bottles to 70°C
Safety pin (to dislodge clogged cDNA bottle tip)

Gloves (1 pair per group)
DNA microarray kit containing:
   10 DNA microarray slides (reusable) (each group needs 1 slide)
   cDNA bottles—one each of #1 (C4BPA), #2 (ODC1), #3 (SIAT9), #4 (FGG), #5 (HBG1), #6 (CYP24) (shared between groups)
   2 hybridization bottles (not heated, shared between groups)
   Alternatively: Micropipets (capacity for 30 μl) and tips

PROCEDURE

Escherichia coli was grown in lactose and in glucose broth. Six known genes were added to your microarray slide. You need to determine which sugar induced gene expression.

Each of the 6 dropper bottles is labeled as 1 (C4BPA), 2 (ODC1), 3 (SIAT9), 4 (FGG), 5 (HBG1), 6 (CYP24) for the 6 genes you will look at today.

1. You will work in groups. Each group will obtain a microarray glass slide by holding the edges of the slide.
2. Add the solution gel from bottle 1 to slide dot 1 on your slide. It should fill the circle but not overflow. Alternatively, you may pipet 30 μl of gel onto your slide. You need a new pipet tip for each dot.
3. You will repeat with the rest of the solution gels onto the slide dots (bottle 2 to dot 2, bottle 3 to dot 3, etc.). Try to make sure to get an equal amount on each dot area.
4. Wait until the dots have hardened before moving onto the next step, usually about 5 minutes.
5. The persons handling the slide and the hybridizing solution must wear gloves.
6. Add 1 or 2 drops of hybridizing solution onto each dot, being careful to not touch the dropper tip to the gel.
7. The dots will change colors. Record the colors you see on each spot in the Laboratory Report.

8. Interpret your results and what they mean. Be sure to record the shade of the color seen (light or dark pink, etc.).

Pink will indicate the *Escherichia coli* genes expressed in glucose.

Blue will indicate *Escherichia coli* genes expressed in lactose.

Purple will indicate *Escherichia coli* genes expressed in cells grown in both sugars.

Colorless will indicate genes are NOT transcribed by cells grown in either sugar.

9. Use a paper towel to wipe off the 6 spots into the small trash can provided. Rinse the slide in water and dry it with another paper towel. Dispose of your gloves and paper towels in the proper container.

10. Place your cleaned slide on a paper towel to dry.

EXERCISE

34

Laboratory Report: Microarray

Results

Indicate the colors you saw for each gene and what sugar induced expression. Explain your results.

Color	Is the gene expressed?	If so, by which sugar?	Result/Conclusions
Dot #1	Yes/No		
#2	Yes/No		
#3	Yes/No		
#4	Yes/No		
#5	Yes/No		
#6	Yes/No		

Questions

1. What would you expect to happen if you added the wrong gene to the slide dot?

2. Would microarrays be more beneficial to science than other technologies like PCR? Why or why not?

INTRODUCTION to the Individual Projects

Many students taking a laboratory class in microbiology are planning careers in medicine and health-related fields. Therefore, the emphasis in many courses is on laboratory exercises that will provide vital skills and concepts to prepare the students for those occupations. The laboratory exercises tend to emphasize the control and identification of pathogenic organisms in addition to an understanding of basic microbiological principles.

Pathogenic organisms, however, make up a very small percentage of the known microorganisms that exist in the world. For the most part, it is the non-pathogenic prokaryotes that are responsible for recycling animal and plant material and are absolutely essential for making life on Earth possible.

How are they able to carry out these activities? Prokaryotes have astonishing versatility. The physiological abilities of prokaryotes make the eukaryotes appear very limited. For instance, no eukaryotes can fix nitrogen, oxidize sulfur for energy, or produce methane, and very few can grow anaerobically. The unique physiological tricks that bacteria use in breaking down and synthesizing molecules as well as their adaptation to their niche are very interesting. For instance, there are organisms that specialize in taking one-carbon compounds produced by plants and building them into cellular material that can then be utilized by other organisms. Others break down complex molecules to simpler components to be incorporated by other bacteria. Some bacteria have such adaptive strategies as producing light or resisting the effects of radiation, which are clearly important to the organisms' survival, although their roles are not understood.

Once you sample the world of microbiology, you may find it especially exciting, and perhaps you would like to investigate an organism on your own as an individual project. The following section presents some protocols for isolating bacteria that are particularly interesting for their unique physiological abilities. Some of these organisms require patience and persistence to isolate. A large part of the project is finding the necessary equipment and source of samples. But if you enjoy a challenge, these individual projects can be very rewarding. Your instructor will decide whether the results should be presented in a written paper, a poster, a report to the class, or both a written and an oral presentation.

Isolating these organisms is also a true investigation. The best source of some of these organisms has not yet been determined. It would be very helpful for future students if the results of these independent investigations were filed in the laboratory, in a separate folder for each organism. You could then add information to the folder about what you learned from your experience with isolating and identifying your particular organism.

Two projects are included in the physical manual on the following pages (Hydrocarbon-Degrading Bacteria and Luminescent Bacteria). Two more are posted online at www.mhhe.com/Nester7e (Methylotrophs and The UV-Resistant *Deinococcus*).

Notes:

35

Hydrocarbon-Degrading Bacteria: Cleaning Up after Oil Spills

Getting Started

The biosphere contains a great variety of organisms that are collectively capable of breaking down just about all naturally occurring carbon compounds. This is fortunate, because otherwise any compound not reduced to simpler molecules would accumulate in the environment. Sometimes it is useful to isolate the specific bacteria that are responsible for breaking down a particular compound. This can be done by using a medium containing only that particular compound as the source of carbon. If you want to isolate an organism, for instance, that oxidizes phenol, you can use a mineral salts medium for the nitrogen, phosphate, and sulfur and then use phenol for the carbon source.

Crude petroleum oil spills are a serious threat to the marine environment. Several methods of removing the oil from the environment were tried in a major oil spill in Alaska. One of the more useful was simply adding fertilizer. It provided a source of nitrogen and phosphorus to encourage the growth of bacteria already present in the environment. In the more recent oil spill in the Gulf of Mexico, the role of these bacteria was carefully studied. Their effect is still being determined (figure 35.1).

The hydrocarbons in oil are natural compounds found almost everywhere in nature. They are made up of a mix of different lengths of carbon chains saturated (covered) with hydrogen, called alkanes, and other carbon–hydrogen compounds. Because petroleum is formed by fossil plants, these same organic molecules are found in garden soil produced by growing plants. The same kinds of organisms that are present in marine environments breaking down seaweed (and potentially petroleum) are also present in the garden soil. Therefore, even though

Figure 35.1 Gulf of Mexico oil spill disaster. Miles of floating barriers are put in place to protect the coastline from a growing oil slick. Source: U.S. Navy.

muddy marine soil is an ideal source for these organisms, garden soil is also an excellent source of hydrocarbon degraders.

Choose any petroleum product—kerosene or fuel oil, for example—for the carbon source. Avoid products containing detergents or other additives.

Enrichment

The first step in these kinds of isolations is enrichment, a way to increase the numbers of the desired bacteria. This means adding inoculum consisting of a source of the organism to a mineral medium with the sole carbon source. In this case, soil will be added to the mineral medium, plus a hydrocarbon. The soil itself contains nutrients that can support the growth of many organisms, but the bacteria that are able to also use the hydrocarbon will have an advantage. They will be able to grow after the carbon compounds added with the soil have been exhausted.

1. For the enrichment, prepare mineral salts medium (MSM) broth (table 35.1). Add 200 ml to a 500 ml flask and cover with a foil cap. Prepare several flasks to maximize your chances of successfully isolating the organisms.

2. Add 0.1% vol/vol of the hydrocarbon of the carbon source (0.2 ml hydrocarbon/200 ml broth). Remember: % vol/vol = ml/100 ml.

3. Add about a gram of soil to the broth.

4. Incubate the flask for 2 weeks at room temperature. These organisms are usually obligate

aerobes, so shaking the flask is helpful but not necessary. Avoid light to discourage the growth of algae. Because they fix CO_2, the algae can grow on CO_2 from the air without degrading the hydrocarbon.

Selection

The second step is selection. After enriching the organism in a broth culture, your chances of isolating it are greatly increased. Streak the broth on mineral salts agar plates and add the hydrocarbon as described. The medium is very selective because only organisms able to utilize the hydrocarbon can grow.

1. Pour mineral salts agar plates. If you use plastic petri plates, check to see if the hydrocarbon you plan to use dissolves them. If it does, use glass plates.

2. Add the hydrocarbon–carbon source, such as fuel oil. (Because some petroleum products do not dissolve well in water, it was not placed in the agar.) Invert the agar plate, and place a piece of filter paper in the lid of the petri dish. Add 0.5 ml of the hydrocarbon to the filter paper, and replace the inverted top of the agar plate (figure 35.2). The organism will be able to grow on the fumes.

3. To obtain isolated colonies, streak the MSM agar plates with a loopful of inoculum from your enrichment flasks.

4. Prepare a control by adding water to the filter paper in another plate. This will control for organisms that might be able to degrade agar to obtain carbon and therefore would grow using agar instead of the hydrocarbon.

5. Incubate the plates at room temperature for about a week.

Table 35.1   Minimal Salts Medium

NaCl	2.5 g (28.4g NaCl for marine organisms)
K_2HPO_4	4.74 g
KH_2PO_4	0.56 g
$MgSO_4 \cdot 7H_2O$	0.50 g
$CaCl_2 \cdot H_2O$	0.1 g
NH_4NO_3	2.5 g
Tap water	1 liter
Agar (for plates)	20 grams (15 grams Difco agar)
pH 7.1	

Note: 2.06 $(NH_4)_2SO_4$ and 3.15 KNO_3 can be substituted for 2.5 g NH_4NO_3.
Source: Modified from E. Rosenberg.

Figure 35.2 The hydrocarbon is placed on the filter paper in the inverted petri dish.

Isolation

1. Examine your plates carefully. Is there more than one type of colony? Note if any organisms are growing on the control plate—these must be using either nutrients contaminating the minerals in the medium or the agar itself (a few bacteria can utilize agar as a carbon source).

2. Purify your isolates by restreaking the organisms on a similar plate with the same hydrocarbon. Repeat until you can be sure you have at least one pure culture.

3. Make a Gram stain for an initial identification of the organism or organisms growing on the plates. If a Gram-negative rod is present, determine if it is oxidase positive. *Pseudomonas* is oxidase positive and well known for its ability to degrade unusual compounds. There are many Gram-positive organisms as well as other Gram-negative bacteria that are also involved in degrading complex molecules. It may not be possible for you to specifically identify your organism, but you can propose possible genera.

4. You might try different hydrocarbons to determine whether your organism can degrade other petroleum products. Does it grow on standard laboratory media? Can you think of other experiments using your isolated organism?

36

Luminescent Bacteria: Bacteria That Produce Light

Getting Started

Some marine bacteria can emit light, a feature called *bioluminescence*. Most of these organisms live in relationship with tropical fishes. In one group of fish, a specialized organ near the eye supports the growth of these bacteria. By covering and uncovering the organ—and so the bacteria—with a special lid, the fish signal other fish in the darkness of deep water. Angler fish have an organ for *luminescent* bacteria that they dangle in front of them like a lure. Smaller fish are attracted to the light and are eaten. In the northern waters, the fish do not have special organs for luminescent bacteria. Here, these bacteria are found in fish intestines, are associated with squid, or are free-living in salt water.

The bacterial enzyme luciferase requires both oxygen and a protein called an inducer. One organism alone cannot produce enough of the inducer to permit light production—an entire population of bacteria is needed. This phenomenon is called *quorum sensing* and has recently been found in many other bacterial systems. Light production is very energy intensive—the equivalent of 6–60 molecules of ATP are needed for each photon of light produced. This system prevents one single bacterium from wasting energy to produce light that cannot be detected.

It is clear that luminescence may be an advantage to bacteria living symbiotically with fish. In exchange for light, they have a protected and rich environment. As previously mentioned, squid also may have special organs associated with these bacteria. Some squid store bacteria in a sac and expel the bacteria when attacked, similar to a cloud of ink. Others may use the bacteria to produce light so that they do not cast a shadow on bright moonlit nights, thereby protecting themselves from predators.

More puzzling is the presence of luminescent bacteria living in the gut of fish. Because this is an anaerobic environment, no light is produced, even though there is sufficient inducer. One proposal, which has some evidence to support it, is that the bacteria prefer the rich environment of the intestine, and when cast out into the water, they try to return by producing light and enticing the fish to swallow them.

Luminescent bacteria are also found free-living in salt water. If they grow on a piece of detritus or organic material, they might accumulate enough inducer to emit light. In this case, a population of bacteria is more successful than a single cell.

Many marine dinoflagellates luminesce when the salt water is warm. They frequently can be seen at night when water is agitated by waves lapping on the beach or by an anchor thrown into the water. It is not known what role luminescence plays in their ecology.

Procedure for Isolating Luminescent Bacteria

1. Prepare several plates of seawater complete agar (table 36.1). No one universal medium supports the growth of all luminescent bacteria, but this has been very successful.

2. Obtain a whole saltwater fish or purchase a squid at a seafood store. Fresh squid is best, but frozen is also good if it is not thawed with hot water. Results with frozen fish have not been very successful.

Table 36.1   Seawater Complete Agar

Seawater	750 ml
Glycerol	3.0 ml
Peptone	5.0 grams
Yeast extract	0.5 grams
Agar	15.0 grams
Water	250 ml
pH 7.5	

Note: The seawater can be actual seawater or any kind of artificial seawater made up in the lab or purchased from an aquarium supply store.
Source: From K. H. Nealson.

3. Dip a swab in the intestinal content of the fish, and swab a third of the agar plate. Similarly, cut the squid open and try swabbing various organs. It is not clear where the luminescent bacteria are in highest concentration, but try swabbing the outside surface and the interior of the squid. Dispose of the swab in a wastebucket, and continue streaking for isolation with a loop.

4. Incubate plates at room temperature for 12–48 hours.

5. Observe for luminescent colonies. Take the plate into a dark room and permit your eyes to adjust for a few minutes. Circle any glowing colonies on the bottom of the plate with a marking pen. Try to look at the plate frequently in the first 24 hours because sometimes the bacteria are luminous for only a few hours. If your first attempt is unsuccessful, try again. Not all squid or fish have luminescent bacteria or at least ones that grow on this medium. If you leave the fish or squid unwrapped at room temperature, you will sometimes see glowing colonies on their surface the next day.

6. Restreak a glowing colony onto another plate to obtain a pure culture. The organism is probably either *Photobacterium* or *Vibrio*.

7. If you would like to demonstrate the luminescence of your organisms to the class, you can write a message or draw an image on an agar plate with a swab dipped in your pure culture.

APPENDIX 1

Culture Media and Reagent Formulas

Dissolve media ingredients and then autoclave media at 121°C for 15 minutes at 15 psi unless otherwise noted.

Blood Agar

Blood agar base (Difco)	40.0 g
Distilled water or deionized water	1000 ml

Dissolve and sterilize. Cool to 45°–50°C. Add 5% sterile, defibrinated sheep blood.

Brilliant Green Lactose Bile Broth

Oxgall, dehydrated	20.0 g
Lactose	10.0 g
Pancreatic digest of gelatin	10.0 g
Brilliant Green	0.013 g

pH 7.2 ± 0.2 at 25°C

Add components to distilled/deionized water and bring volume to 1.0 L. Mix thoroughly. Distribute into tubes containing inverted Durham tubes, in 10.0 ml amounts for testing 1.0 ml or less of sample. Autoclave for 12 min. After sterilization, cool the broth rapidly. Medium is sensitive to light.

Cytophaga Agar (10 ml/screw cap tube)

Per liter: Tryptone	0.5 g
Yeast extract	0.5 g
Beef extract	0.2 g
Sodium acetate	0.2 g
Agar	4.0 g

pH 7.2 before sterilizing
pH 7.2–7.3 after sterilizing

Corn Meal Agar

Per liter: Corn meal	60.0 g
Agar	15.0 g
Tween 80*	10.0 g

Mix the corn meal to a smooth cream with water, simmer for one hour, filter through cheesecloth or coarse filter paper, add the agar and Tween 80 and heat until dissolved, make up to volume, and sterilize, preferably for 30 minutes at 10 pounds pressure.

*Tween 80 addition greatly stimulates chlamydospore formation.

Endo Broth/Agar

Per liter: Peptone	10.0 g
Lactose	100.0 g
KH_2PO_4	3.5 g
Sodium sulfate	2.5 g
Agar (leave out for broth)	10.0 g

pH 7.5 (approx.)

Add components to distilled water. Add 4 ml of a 10% (weight/volume) alcoholic solution of basic fuchsin (95% ethyl alcohol). Bring to a boil to dissolve completely. Mix well before pouring.

Levine Eosin Methylene Blue (EMB) Agar

Per liter: Peptone	10.0 g
Lactose	10.0 g
K_2HPO_4	2.0 g
Eosin Y	0.4 g
Methylene blue	0.065 g
Agar	15.0 g

Add components to distilled water. Boil to dissolve completely. Sterilize at 121°C for 15 minutes. Shake medium at 60°C to oxidize the methylene blue (i.e., restore its blue color) and suspend the precipitate, which is an important part of the medium.

Glucose–Acetate Yeast Sporulation Agar

Per liter: Glucose	1.0 g
Yeast extract	2.0 g
Sodium acetate (with 3 H_2O)	5.0 g
Agar	15.0 g

Adjust pH to 5.5

Halobacterium Medium (ATCC Medium 213)

NaCl	250 g
$MgSO_4 \cdot 7H_2O$	10 g
KCl	5 g
$CaCl2 \cdot 6H_2O$	0.2 g
Yeast extract	10 g
Tryptone	2.5 g
Agar	20 g

The quantities given are for 1 L final volume. Make up two solutions, one containing yeast extract, tryptone, and agar, and the other the salts. Adjust pH of the agar solution to 7.0. Sterilize the solutions separately at 121°C for 15 minutes. Mix and dispense aseptically approximately 6 ml per 13 mm sterile tube.

Lauryl Sulfate Broth (Double Strength)
Halve to Make Single Strength

Per liter: Tryptose	40.0 g
Lactose	10.0 g
NaCl	10.0 g
K_2HPO_4	5.5 g
KH_2PO_4	5.5 g
Sodium lauryl sulfate	0.2 g

pH 6.8 (approx.)

Dispense 10 ml quantities in large test tubes containing inverted Durham tubes.

MacConkey Agar

Per liter: Peptone	17.0 g
Proteose peptone	3.0 g
Bacto bile salts no. 3	10.0 g
NaCl	5.0 g
Agar	13.5 g

Neutral red   0.03 g
Crystal violet   0.001 g
pH 7.1

Mannitol Salt Agar for *Staphylococcus* Isolation

Phenol red agar base	1,000 ml
d-mannitol	10.0 g
NaCl	75.0 g

Add the d-mannitol and sodium chloride to the phenol red base while it is still melted, and heat to dissolve.

Methyl Red-Voges–Proskauer Medium

Per liter: Buffered peptone	7.0 g
Dipotassium phosphate	5.0 g
Glucose	5.0 g

Final pH 6.9 ± 0.2
Distribute 5 ml/tube
Two tubes are needed for the test, one for Voges–Proskauer and the other for methyl red.

Mineral Salts + 0.5% Glucose (or Other Sugar)

Per liter:	*1×*	*10×*
$(NH_4)_2SO_4$	1.0 g	10.0 g
K_2HPO_4	7.0 g	70.0 g
KH_2PO_4	3.0 g	30.0 g
$MgSO_4 \cdot 7\,H_2O$	0.1 g	1.0 g

pH 7.0
Sterilize 20 minutes at 121°C.

Glucose agar base:

Agar	6.0 g
Glucose	2.0 g
Distilled water	360 ml

Sterilize for 15 minutes at 118°C.
Add 40 ml of the 10× salts to 360 ml glucose agar before pouring into plates.

Motility Medium

Per liter: Dehydrated nutrient broth	8.0 g
Agar	3.0 g

Mix the dehydrated nutrient broth and agar in the water. Heat until the agar has dissolved and adjust to pH 6.8.
Dissolve 25 mg TTC in 5 ml of distilled water. Make up the dye solution just before it is added to the medium.

Triphenyl tetrazolium chloride (TTC)	5 ml of 0.5% aqueous solution

Add the diluted dye. Dispense 10 ml/tube.

Mueller–Hinton Agar

Per liter: Beef, infusion from	300.0 g
Casamino acids, tech	17.5 g
Starch	1.5 g
Agar	17.0 g
When using a commercial base: Mueller–Hinton broth (dehydrated)	21.0 g
Agar	17.0 g

Nutrient Broth/Agar

Per liter: Beef extract	3.0 g
Peptone	5.0 g
Agar (leave out for broth)	15.0 g

pH 6.9 (approx.)

Nutrient + Starch Agar

Nutrient broth	1,000 ml
Agar	15.0 g
Soluble starch	2.0 g

Prepare nutrient broth. Add agar and heat to melt agar. Suspend the starch in a small quantity of cold water and add to the nutrient agar while it is still hot. pH 7 before and after sterilizing.

Nutrient + Starch + Glucose Agar

Add 20 grams (2%) glucose to above medium.

1% Peptone Agar Deeps (10ml/screw cap tube)

Per liter: Peptone	10.0 g
Agar	10.0 g

Sabouraud's Dextrose Broth/Agar

Per liter: Dextrose	40.0 g
Peptone	10.0 g
Agar (leave out for broth)	15.0 g

Final pH 5.6 (approx.).

Saline (Per liter)

NaCl	8.5 g
Distilled water	1,000 ml

Simmons Citrate Agar

Per liter: Magnesium sulfate	0.2 g
Monoammonium phosphate	1.0 g
Dipotassium phosphate	1.0 g
Sodium citrate	2.0 g
NaCl	5.0 g
Agar	15.0 g
Bromthymol blue	0.08 g

pH 6.8. Distribute in tubes or flasks. Cool in a slanting position.

Sugar Fermentation Tubes (Andrade's)

Per liter: Nutrient broth (dehydrated)	8.0 g
Glucose or other sugar	5.0 g
Andrade's indicator*	10 ml

pH 7.1 before sterilizing, pH 7.0 after sterilizing
Add 6 ml/standard tube. Add Durham tubes inverted before sterilizing—they will fill in the autoclave. Sterilize for 12–13 minutes at 118°C.

* Andrade's indicator:

Acid fuchsin	1.0 g
Distilled water	750 ml
1 M NaOH	7 ml

To dissolve and neutralize the acid fuchsin solution, add 7 ml 1M NaOH. Allow to stand for several hours. If the solution is still red, add 3–4 ml more of 1M NaOH. Let stand for several hours. If solution is still red, add 1–2 ml more. Continue until solution is light red. Let stand for several hours and adjust pH to 8.0—it will still appear a light red. (When added to medium, the broth will be a very pale yellow.) This can be stored for long periods at room temperature.

Sugar Fermentation Tubes (Phenol Red)

Per liter: Beef extract	1.0 g
Proteose peptone no. 3	10.0 g

NaCl	5.0 g
Phenol red	0.018 g

Final pH 7.4

Dissolve 16 g phenol red broth base in 100 ml of distilled water. Add 5 g of carbohydrate. Add 6 ml/standard tube. Add inverted Durham tubes. Test the first batch with known fermenters. Occasionally the carbohydrate hydrolyzes. If this occurs, lower temperatures and shorter sterilization time are recommended, testing for sterility.

Thioglycollate Medium, Fluid

Per liter: Yeast extract	5.0 g
Casitone	15.0 g
Dextrose	5.0 g
NaCl	2.5 g
1-Cystine	0.75 g
Thioglycollic acid	0.3 ml
Agar	0.75 g
Resazurin, certified	0.001 g

Final pH 7.1. Suspend 29.5g of the thioglycollate medium in 1000 ml cold distilled water and heat to boiling to dissolve completely. Distribute into tubes and sterilize for 15 minutes at 121 degrees C.

Tryptone Broth

Per liter: Tryptone	10.0 g
NaCl	5.0 g

pH 7.0 before sterilizing

Tryptone Agar for Base Plates

Per liter: Tryptone	10.0 g
NaCl	5.0 g
$MgSO_4 \bullet 7H_2O$	1.0 g
Agar	15.0 g

pH 7 before sterilizing.

Tryptone Soft Agar Overlay

Same formula as above but use 7 grams of agar.

Trypticase (Tryptic) Soy Broth (TSB)/Agar

Per liter: Trypticase peptone	17.0 g
Phytone peptone	3.0 g
NaCl	5.0 g
K_2HPO_4	2.5 g
Glucose	2.5 g
Agar (leave out for broth)	15.0 g

pH 7.3 (approx.)

For TS + streptomycin plates: Add 1.0 ml of stock streptomycin per liter of sterile, melted, cooled agar (stock streptomycin: 1.0 g streptomycin/10 ml distilled water, filter sterilized).

Trypticase Soy Yeast (TSY) Extract Agar

Per liter: Trypticase soy broth	30.0 g
Yeast extract	10.0 g
Agar	15.0 g

pH 7.2 before sterilizing

TSY + Glucose Agar

Add 2.5 grams glucose/liter.

TSY + Glucose + Brom Cresol Purple Agar Slants

Per liter: Trypticase soy broth	30.0 g
Yeast extract	10.0 g
Glucose	7.5 g
Brom cresol purple	0.04 g

Agar	15.0 g

pH 7.3 before sterilizing, pH 7.0 after sterilizing. Sterilize at 118°C for 15 minutes, 6 ml/tube.

TYEG Salts Agar Plates Containing 0.5, 5, 10, and 20% NaCl

For basal medium composition see TYEG salts agar. The two lower salt concentrations (0.5 and 5%) can be added directly to the basal medium, but the two higher salt concentrations (10 and 20%) need to be autoclaved separately and added back to the basal medium after autoclaving. For related procedural details see the ATCC agar medium 217 procedure.

TYEG Salts Agar Plates Containing 0, 10, 25, and 50% Sucrose

For basal medium composition see TYEG salts agar. The three 10% sucrose concentrations can be added directly to the basal medium, whereas the remaining two sucrose concentrations (25% and 50%) need to be autoclaved separately and added back to the basal medium after autoclaving. For related procedural details see the ATCC agar medium 213 procedure.

TYEG (Tryptone Yeast Extract Glucose) Salts Agar Slants

Per liter: Tryptone	10 g
Yeast extract	5 g
Glucose	2 g
K_2HPO_4	3 g
$CaCl_2 \bullet 6H_2O$	0.2 g
Agar	20 g

pH 7.0 (before autoclaving). Dispense 6 ml/tube.

Urea Broth/Agar

Urea broth/agar base	29.0 g

Ingredients in this base:

Peptone	1.0 g
Dextrose	1.0 g
NaCl	5.0 g
Monopotassium Phosphate	2.0 g
Urea	20.0 g
Phenol red	0.012 g
Agar (leave out for broth)	15.0 g

Final pH 6.8

Suspend 29 g of urea agar base in 100 ml of distilled water. Filter sterilize this concentrated base. Dissolve 15 g of agar in 900 ml distilled water by boiling, and sterilize. Cool to 50° to 55°C and add 100 ml of the filter sterilized urea agar base under aseptic conditions. Mix and distribute in sterile tubes. Slant tubes so they will have a butt of 1 inch and slant 1½ inches in length, and solidify in this position.

Worfel–Ferguson Slants

Per liter: NaCl	2.0 g
K_2SO_4	1.0 g
$MgSO_4 \bullet 7H_2O$	0.25 g
Sucrose	20.0 g
Yeast extract	2.0 g
Agar	15.0 g

No pH adjustment.

Yeast Fermentation Broth Containing Durham Tubes

Per liter: Tryptone	10.0 g
Yeast extract	5.0 g
K_2HPO_4	5.0 g
Sugar (glucose, lactose, or maltose)	10.0 g

The method of preparation is the same for phenol red broth base.

Note: Be careful not to overautoclave, because the sugars may decompose with overheating.

Acid Alcohol for Acid-Fast Stain

Ethanol (95%)	97.0 ml
Concentrated HCl	3.0 ml

Add acid to alcohol. Do not breathe acid fumes.

Carbolfuchsin Stain

Basic fuchsin	0.3 g
Ethanol (95%)	10.0 ml
Phenol	5.0 ml
Distilled water	100.0 ml

Dissolve the basic fuchsin in ethanol. Add the phenol to the water. Mix the solutions together. Allow to stand for several days. Filter before use.

DNase (Deoxyribonuclease)

Sigma DN-25 or any crude grade
Add a few visible crystals to 5 ml sterile water.
Hold in ice bucket when in solution.

Gram-Stain Reagents

Crystal Violet (Hucker Modification)

Solution A:	Crystal violet (gentian violet)	40.0 g
	Ethanol (95%)	400.0 ml
Solution B:	Ammonium oxalate	1.0 g
	Distilled water	1,600.0 ml

Dissolve the crystal violet in the alcohol and let stand overnight. Strain through filter paper. Add solution B. Mix to dissolve completely.

Gram's Iodine

Iodine	1.0 g
Kl	2.0 g
Distilled water	300.0 ml

Dissolve the Kl in 2 ml of water. Add the iodine and dissolve. Then add the 300 ml of water.

Decolorizer for Gram Stain

Use either 95% ethanol or ethanol/acetone 1:1.

Safranin O (Distilled water)

Safranin	10.0 g

Hydrogen Peroxide (3%)

Purchase 3% H_2O_2 and put in brown dropper bottles. Refrigeration is recommended.

India Ink

Smaller particles are better than larger particles.

Kovacs Reagent

Isoamyl alcohol (isopentyl)	75.0 ml
Concentrated HCl	25.0 ml
p-dimethylaminobenzaldehyde	5.0 g

Mix the alcohol and the acid. Add the aldehyde and stir to dissolve. Store in plastic or dark glass bottle.

Malachite Green (5%)

Malachite green	5.0 g
Distilled water	95.0 ml

Mix, filter, and store in brown bottle.

Methyl Red

Methyl red	0.1 g
Ethanol (95%)	300.0 ml
Distilled water	200.0 ml

Dissolve methyl red in alcohol and add water. Dispense into dropper bottles.

Methylene Blue (Loeffler's)

Solution A:	Methylene blue	0.3 g
	95% Ethanol	30.0 ml
Solution B:	0.01% (w/v) KOH	100.0 ml
	about 1 pellet KOH/	1,000 ml

Mix solutions A and B together.

Methylene Blue (Acidified)

Solution A:	Methylene blue	0.02 g
	Distilled water	100.0 ml
Solution B:	KH_2PO_4 (0.2 Molar)	99.75 ml
Solution C:	K_2HPO_4 (0.2 Molar)	0.25 ml

Mix solutions A, B, and C together (pH must be 4.6).

ONPG

Per liter: O-nitrophenyl β-D galactopyranoside	1.0 g (.0003 M)

Oxidase Reagent

N,N-dimethyl-p-phenylene-diamine hydrochloride	1.0 g
Distilled water	100.0 ml
pH 5.5	

This must be prepared on the day it is used. However, it can be frozen in small vials and thawed when needed, but the test is more difficult to read.

Phosphate Buffer

Per liter: $Na_2HPO_4 \cdot 2H_2O$	11.88 g
KH_2PO_4	9.08 g

For pH 6.8: mix 6 parts Na_2HPO_4 with 4 parts KH_2PO_4.

Sodium Dodecyl Sulfate (SDS) in 10× Saline Citrate

Per liter: Sodium dodecyl sulfate	5.0 g
NaCl	87.7 g
Sodium citrate	44.1 g

No pH adjustment.

Streptomycin

Prepare a stock solution of Streptomycin sulfate by adding 1 g to 10 ml distilled water. Filter sterilize, and freeze in convenient amounts.

Voges–Proskauer Reagent

Reagent A:	Alpha-naphthol	5.0 g
	Ethyl alcohol (95%)	100 ml (bring to volume)
Reagent B:	KOH	40 g
	Distilled water	100 ml (bring to volume)

Dispense each reagent in separate dropper bottles.

2

Living Microorganisms (Bacteria, Fungi, Protozoa, and Helminths) Chosen for Study in This Manual

Acinetobacter, aerobic Gram-negative rods or coccobacilli in pairs. Commonly found in soil. Low virulence, opportunistic pathogen. Naturally competent and therefore easily transformed.

Alcaligenes faecalis, a Gram-negative rod, obligate aerobe, nonfermentative, found in the intestines.

Amoeba proteus, a unicellular protozoan that moves by extending pseudopodia.

Aspergillus niger, a filamentous black fungus with a foot cell, columella, and conidia.

Bacillus cereus, a Gram-positive rod, forms endospores, found in soil.

Bacillus subtilis, a Gram-positive rod, forms endospores, found in soil.

Citrobacter freundii, a Gram-negative rod, produces sulfur, found in the intestinal tract and soil.

Clostridium sporogenes, a Gram-positive rod, forms endospores, obligate anaerobe, found in soil.

Diphtheroids, Gram-positive irregular club-shaped rods.

Dugesia, a free-living flatworm.

Enterobacter aerogenes, a Gram-negative rod, coliform group, found in soil and water.

Enterococcus faecalis, a Gram-positive coccus, grows in chains, found in the intestinal tract; opportunistic pathogen.

Escherichia coli, a Gram-negative rod, facultative, found in the intestinal tract coliform group, can cause diarrhea and serious kidney disease.

Escherichia coli, K-12 strain, commonly used in research. Host strain for λ phage.

Halobacterium salinarium, member of the Archaea, can live only in high salt solutions.

Klebsiella pneumoniae, Gram-negative rod, coliform group, opportunistic pathogen.

Micrococcus luteus, Gram-positive obligate aerobe, cocci are arranged in packets of four or eight. Part of the normal biota of the skin. The yellow colonies are frequently seen as an air contaminant.

Moraxella, Gram-negative coccus found in normal biota.

Mycobacterium phlei, an acid-fast rod with a Gram-positive type of cell wall.

Mycobacterium smegmatis, an acid-fast rod with a Gram-positive type of cell wall.

Paramecium, a ciliated protozoan found in ponds.

Penicillium species, a filamentous fungus with metulae, sterigmata, and conidia; makes antibiotics.

Propionibacterium acnes, Gram-positive irregular rod, obligate anaerobe, component of the normal skin biota living in sebaceous glands.

Proteus, a Gram-negative rod, swarms on agar, hydrolyzes urea, can cause urinary tract infections.

Pseudomonas aeruginosa, a Gram-negative rod, obligate aerobe, motile, opportunistic pathogen, can degrade a wide variety of compounds.

Rhizopus nigricans, a filamentous fungus with stolons, coenocytic hyphae, and a sporangium containing asexual sporangiospores.

Saccharomyces cerevisiae, eukaryotic fungal yeast cell, replicates by budding. Important in bread, beer, and wine making and in fungal genetics.

Salmonella enteriditis, a Gram-negative rod, nonfermentative, produces sulfur, found in the intestines.

Serratia marcescens, a Gram-negative rod that produces pigments, found in soil.

Shigella sonnei, a Gram-negative rod, nonfermentative, found in the intestines.

Spirillum volutans, a Gram-negative curved rod, flagella at each pole, found in pond water.

Staphylococcus aureus, a Gram-positive coccus, a component of the normal skin biota, but can cause wound infections and food poisoning.

Staphylococcus epidermidis, a Gram-positive coccus, a component of the normal skin biota.

Streptococcus bovis, a Gram-positive coccus, a non-enterococcus, found in soil.

Streptococcus mutans, normal biota of the mouth, forms gummy colonies when growing on sucrose.

Streptococcus pneumoniae, lancet-shaped Gram-positive cells arranged in pairs and short chains, pathogenic strains form capsules.

Streptococcus pyogenes, Gram-positive cocci in chains, the cause of strep throat.

3

Dilution Practice Problems

See exercise 8 for an explanation of making and using dilutions.

1. If a broth contained 4.3×10^2 org/ml, about how many colonies would you expect to count if you plated
   a. 1.0 ml?
   b. 0.1 ml?

2. Show three ways for making each
   a. 1/100 or 10^{-2} dilution,
   b. 1/10 or 10^{-1} dilution, and
   c. 1/5 or 2×10^{-1} dilution.

3. Show two ways of obtaining a 10^{-3} dilution using 9.0 ml and 9.9 ml dilution blanks.

4. The diagram below shows a scheme for diluting yogurt before making plate counts. 0.1 ml was plated on duplicate plates from tubes B, C, and D. The numbers in the circles represent plate counts after incubation.

a. Which plates were in the correct range for accurate counting?
b. What is the average of the plates?
c. What is the total dilution of the tubes?:

A _____

B _____

C _____

D _____

d. How many organisms/ml were in the original sample of yogurt?

5. Suppose an overnight culture of *E. coli* has 2×10^9 cells/ml. How would you dilute it so that you have countable plates? Diagram the scheme.

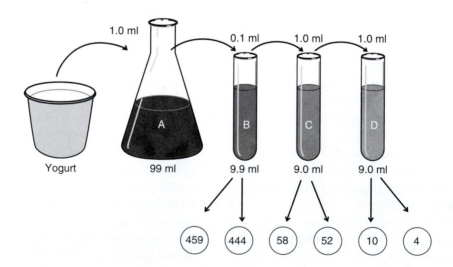

Answers

1a. 430 colonies
1b. 43 colonies
2a. 1.0 ml into 99 ml
    0.1 ml into 9.9 ml
    10 ml into 990 ml
2b. 1.0 ml into 9.0 ml
    0.1 ml into 0.9 ml
    10 ml into 90 ml
2c. 1.0 ml into 4 ml
    0.1 ml into 0.4 ml
    10 ml into 40 ml
3.

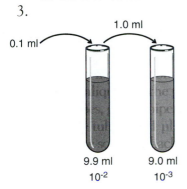

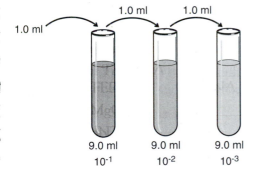

4a. Plates from tube C
4b. 55
4c. A. A: 10^{-2}
    B. B: 10^{-4}
    C. C: 10^{-5}
    D. D: 10^{-6}
4d. The number of colonies $\times$; 1/dilution $\times$;
    1/sample on plate = number of organisms/ml
    55 $\times$; $1/10^{-5}$ $\times$; 1/0.1 = 55 $\times$; 10^{5} $\times$; 10 =
    55 $\times$; 10^{6} or 5.5 $\times$; 10^{7} org/ml
5.

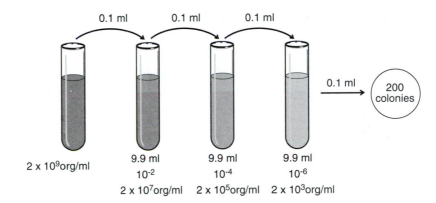

APPENDIX

4

A. Spread Plate Technique

An alternative to making plate counts as described in exercise 8 is the spread plate technique (figure 1). Instead of putting the sample in melted agar, it is placed on the surface of an agar plate and spread around with a bent glass rod (which resembles a hockey stick). Usually the sample size is 0.1 ml, so the following is a dilution scheme for suspension A and suspension B.

First Session

Suspension A

1. Label all water blanks with the dilution:
   10^{-2}, 10^{-3}, 10^{-4}, 10^{-5}, 10^{-6}, 10^{-7}, 10^{-8}. Also label four petri dishes (on the bottom):
   10^{-5}, 10^{-6}, 10^{-7}, 10^{-8}.

2. Make serial dilutions of the bacterial suspension A:
   a. Mix the bacterial suspension by rotating between the hands, and transfer 1.0 ml of the suspension to the 99 ml water blank labeled 10^{-2}. Discard the pipet.
   b. Mix well, and transfer 1.0 ml of the 10^{-2} dilution to the 9.0 ml water blank labeled 10^{-3}. Discard the pipet.
   c. Mix well, and transfer 1.0 ml of the 10^{-3} dilution to the 9.0 ml water blank labeled 10^{-4}. Discard the pipet.
   d. Mix well, and transfer 1.0 ml of the 10^{-4} dilution to the 9.0 ml water blank labeled 10^{-5}. Discard the pipet.
   e. Mix well, and transfer 1.0 ml of the 10^{-5} dilution to the 9.0 ml water blank labeled 10^{-6}. Discard the pipet.

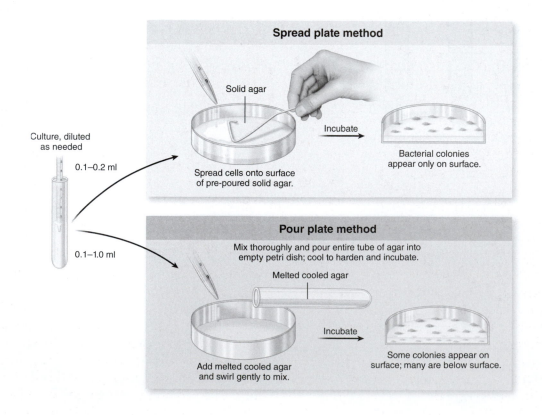

Culture, diluted as needed

0.1–0.2 ml

0.1–1.0 ml

Spread plate method

Solid agar

Incubate

Spread cells onto surface of pre-poured solid agar.

Bacterial colonies appear only on surface.

Pour plate method

Mix thoroughly and pour entire tube of agar into empty petri dish; cool to harden and incubate.

Melted cooled agar

Incubate

Add melted cooled agar and swirl gently to mix.

Some colonies appear on surface; many are below surface.

f. Mix well, and transfer 1.0 ml of the 10^{-6} dilution to the 9.0 ml water blank labeled 10^{-7}. Discard the pipet.

g. Mix well, and transfer 1.0 ml of the 10^{-7} dilution to the 9.0 ml water blank labeled 10^{-8}. Discard the pipet.

h. Mix well.

3. Starting with the most dilute suspension (10^{-8}), remove 0.1 ml and place on the surface of the agar plate. Immediately spread it around the entire surface of the plate with a sterile bent glass rod.

4. Using the same pipet and sterile glass rod (be careful not to contaminate), repeat the procedure for the 10^{-7}, 10^{-6}, and 10^{-5} dilutions.

5. Discard the pipet, and place the glass rod in a container as indicated by the instructor.

6. Invert the plates and incubate at 37°C.

Suspension B

1. Label all water blanks with the dilution: 10^{-2}, 10^{-3}, 10^{-4}, 10^{-5}, 10^{-6}. Also label four petri dishes (on the bottom): 10^{-3}, 10^{-4}, 10^{-5}, 10^{-6}.

2. Make serial dilutions of the bacterial suspension B:

a. Mix the bacterial suspension by rotating between the hands, and transfer 1.0 ml of the suspension to the 99 ml water blank labeled 10^{-2}. Discard the pipet.

b. Mix well, and transfer 1.0 ml of the 10^{-2} dilution to the 9.0 ml water blank labeled 10^{-3}. Discard the pipet.

c. Mix well, and transfer 1.0 ml of the 10^{-3} dilution to the 9.0 ml water blank labeled 10^{-4}. Discard the pipet.

d. Mix well, and transfer 1.0 ml of the 10^{-4} dilution to the 9.0 ml water blank labeled 10^{-5}. Discard the pipet.

e. Mix well, and transfer 1.0 ml of the 10^{-5} dilution to the 9.0 ml water blank labeled 10^{-6}. Discard the pipet.

3. Starting with the most dilute suspension (10^{-6}), remove 0.1 ml and place on the surface of the agar plate. Immediately spread it around the entire surface of the plate with a sterile bent glass rod.

4. Using the same pipet and sterile glass rod (be careful not to contaminate), repeat the procedure for the 10^{-5}, 10^{-4}, and 10^{-3} dilutions.

5. Discard the pipet, and place the glass rod in a container as indicated by the instructor.

6. Invert the plates and incubate at 37°C.

Second Session for Both A and B Suspension (See pages 56–57)

Note: In the spread plate procedure, all the samples are 0.1 ml. This appears as $\frac{1}{0.1}$ for the sample size in the formula:

$$\text{number of organisms/ml in original sample} = \text{the number of colonies on plate} \times \frac{1}{0.1} \times \frac{1}{\text{dilution}}$$

B. An alternate method of calculating concentrations of organisms from dilution plate counts.

1. Convert the dilution to the dilution factor, which is the reciprocal of the dilution. For example, the dilution factor of a 10^{-1} dilution is 10. Convert all dilutions and samples to dilution factors, and then multiply by the average number of colonies on a countable plate.

2. For example: On page 55, the total dilution of the tubes in the box would be $10 \times 10 = 100$.

   If a 1.0 ml sample from tube B contained 48 bacteria, the concentration in the original solution (not shown) would be $48 \times 10 \times 10$ organisms/ml or 48×100 organisms/ml.

3. Using the same example, if 0.1 ml were sampled, the reciprocal of 1/.1 is 10. The concentration would be $10 \times 10 \times 10 \times 48$ organisms/ml or 48×10^3 organisms/ml.

C. An alternative test for coagulase. The slide coagulase test is described in exercise 22. An alternative tube coagulase test involves the following steps. Mix a loopful of cells with 0.5 ml undiluted rabbit plasma in a small tube. Incubate at 37°C, and examine after 4–24 hours. The plasma can also be diluted 1:4 with saline. A solid clot is a positive test. This requires more reagent and takes longer, but some find it easier to read (color plate 26).

Use of the Ocular Micrometer for Measurement of Relative and Absolute Cell Size

Determination of cell dimensions is often used in microbiology, where it has numerous applications. Examples include measuring changes in cell size during the growth cycle, determining the effect of various growth factors on cell size, and serving as a taxonomic assist in culture identification. Measurements are made by inserting a glass disc with inscribed graduations (figure 1a), called an ocular micrometer, into the ocular of the microscope (figure 2).

It is not necessary to calibrate the ocular micrometer to determine the *relative* size of cells for this purpose, either a wet mount or stained preparation of cells can be examined with the ocular micrometer. By measuring the length in terms of number of ocular micrometer divisions, you might conclude that cell X is twice as long as cell Y. For such purposes, determination of cell length in absolute terms—such as number of micrometers—is not necessary.

For determining the *absolute* size of cells, it is necessary to first measure the length in micrometers (μm) between two lines of the ocular micrometer. A stage micrometer with a scale measured in micrometers is needed for μm calibration. The stage micrometer scale (figure 1b) is such that the distance between two lines is 0.01 mm (equivalent to 10 μm). By superimposing the stage micrometer scale over the ocular micrometer scale, you can determine the absolute values in microns between two lines on the ocular micrometer scale. The absolute value obtained is also dependent on the objective lens used. For example, with the low-power objective lens, seven divisions on the ocular micrometer equal one division on the stage micrometer. Thus, with the low-power objective lens, one division on the ocular micrometer equals .01 mm/7, which equals 1.40 micrometers (μm).

Procedure for Insertion, Calibration, and Use of the Ocular Micrometer

1. Place a clean stage micrometer in the mechanical slide holder of the microscope stage.
2. Using the low-power objective lens, center and focus the stage micrometer.
3. Unscrew the top lens of the ocular to be used, and then carefully place the ocular micrometer with the engraved side down on the diaphragm inside the eyepiece tube (figure 2). Replace the top lens of the ocular.
   Note: With some microscopes, the ocular micrometer is inserted in a retaining ring located at the base of the ocular.
4. To calibrate the ocular micrometer, rotate the ocular until the lines are superimposed over the lines of the stage micrometer.
5. Next, move the stage micrometer until the lines of the ocular and stage micrometer coincide at one end.
6. Now find a line on the ocular micrometer that coincides precisely with a line on the stage micrometer.
7. Determine the number of ocular micrometer divisions and stage micrometer divisions where the two lines coincide. An example of this step is shown in figure 1c.

A relatively large practice microorganism for determining average cell size, with both the low- and high-power objective lenses, is a yeast cell wet mount. You must first calibrate the ocular micrometer scale for both objective lenses, using the stage micrometer. Next determine the average cell size of a group of cells, in ocular micrometer units, using both the low- and high-power objective

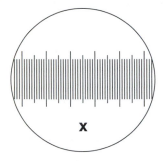

(a) Ocular micrometer
The diameter (width) of the graduations in microns must be determined for each objective.

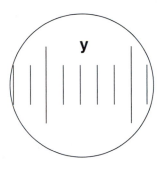

(b) Stage micrometer
The graduations are .01 mm (10 mm) wide.

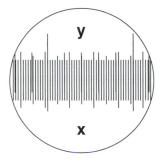

(c) Superimposing of ocular micrometer scale (x) over the stage micrometer scale (y)
Note that seven divisions of the ocular micrometer equal one division of the stage micrometer scale (.01 mm).

One division of x = $\dfrac{.01}{7}$

= .0014 mm = 1.4 mm

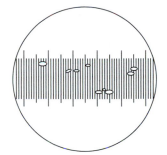

(d) Based on the Figure 2c calculations, what is the average length in microns of the rod-shaped bacteria?

Figure 1 Calibration of the ocular micrometer.

lenses. Attempt to measure the same group of cells with both lenses. You may find it easier to first measure them with the high-power objective lens, followed by the low-power objective lens. You will have mastered the technique if you obtain the same average cell size answer with both objective lenses.

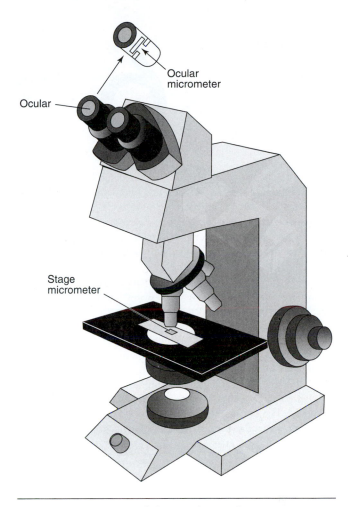

Figure 2 Location of the ocular and stage micrometers.

Note: Care should be taken when inserting and removing the ocular micrometer from the ocular. Before inserting, make certain the ocular micrometer is free of dust particles by cleaning both sides with lens paper moistened with a drop of lens cleaning solution. Install and remove only in an area free of air currents. After removing the ocular micrometer, reexamine it for dust particles. If present consult your instructor.

APPENDIX 6

History and Working Principles of Light Microscopy

History

Anton van Leeuwenhoek (1632–1723), a Dutch linen draper, recorded the first observations of living microorganisms. He used a homemade microscope containing a single glass lens powerful enough to enable him to see what he described as little "animalcules" (now known as bacteria) in scrapings from his teeth and larger "animalcules" (now known as protozoa and algae) in droplets of pond water and hay infusions.

A single-lens microscope such as van Leeuwenhoek's had many disadvantages. These included distortion with increasing magnifying powers and a decreasing focal length (the distance between the specimen when in focus and the tip of the lens). Thus, when using a single lens with an increased magnifying power, van Leeuwenhoek had to practically push his eye into the lens in order to see anything.

Modern microscopes, called **compound microscopes,** have two lenses: an eyepiece or ocular lens and an objective lens (figure 1). The eyepiece lens allows comfortable viewing of the specimen from a distance. It also has some magnification capability, usually 10 times (10×) or 20 times (20×). The purpose of the objective lens, which is located near the specimen, is to provide image magnification and clarity. Most teaching microscopes have three objective lenses with different powers of magnification (usually 10×, 45×, and 100×). Total magnification is obtained by multiplying the magnification of the ocular lens by the magnification of the objective lens. Thus, when using a 10× ocular lens with a 45× objective lens, the total magnification of the specimen image is 450 diameters.

Another giant in the early development of the microscope was a German physicist, Ernst Abbe, who developed (ca. 1883) various microscope improvements. One was the addition of a third lens, the **condenser lens,** which is located below the microscope stage (figure 1). By moving this lens up or down, it is possible to concentrate (intensify) the light emanating from the light source on the bottom side of the specimen slide. (The specimen is located on the top surface of the slide.)

Abbe also developed the technique of using lens immersion oil in place of water as a medium for transmission of light rays from the specimen to the lens of the **oil immersion objective lens**. Oil with a density more akin to the microscope lens than that of water helps to decrease the loss of transmitted light, which, in turn, increases image clarity. Finally, Abbe developed improved microscope objective lenses that were able to reduce both **chromatic** and **spherical lens aberrations**. His objective lenses had a concave lens (glass bent inward like a dish) in addition to the basic convex lens (glass bent outward). Such a combination diverges the peripheral rays of light only slightly to form an almost flat image. The earlier simple convex lenses produced distorted image shapes due to spherical lens aberrations and distorted image colors due to chromatic lens aberrations.

Spherical Lens Aberrations   These occur because light rays passing through the edge of a convex lens are bent more than light rays passing through the center. The simplest correction is the placement of a diaphragm below the lens so that only the center of the lens is used (locate the **iris diaphragm** in figure 1). Such aberrations can also be corrected by grinding the lenses in special ways.

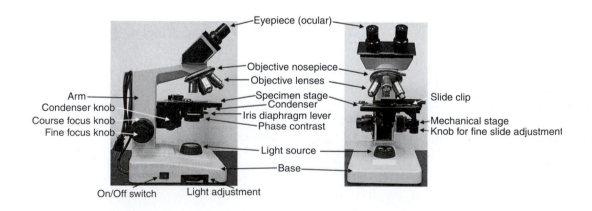

Figure 1 A modern bright-field microscope. Courtesy of Anna Oller, University of Central Missouri.

Chromatic Lens Aberrations These occur because light is refracted (bent) as well as dispersed by a lens. The blue components of light are bent more than the red components. Consequently, the blue light travels a shorter distance through the lens before converging to form a blue image. The red components travel a longer distance before converging to form a red image. When these two images are seen in front view, the central area, in which all the colors are superimposed, maintains a white appearance. The red image, which is larger than the blue image, projects beyond the central area, forming red edges outside of the central white image. Correction of a chromatic aberration is much more difficult than correction of a spherical aberration because dispersion differs among kinds of glass. Objective lenses free of spherical and chromatic aberrations, known as **apochromatic objectives,** are now available but are also considerably more expensive than **achromatic objectives.**

Some Working Principles of Light Microscopy

Microscope Objectives: The Heart of the Microscope

All the other parts of the microscope are involved in helping the objective lens attain a noteworthy image. Such an image is not necessarily the *largest*, but it is the *clearest*. A clear image helps achieve a better understanding of specimen structure. Size alone does not help achieve this end. The ability of the microscope to reveal specimen structure is termed **resolution,** whereas the ability of the microscope to increase specimen size is termed **magnification.**

Resolution, or resolving power, is also defined as the ability of an objective lens to distinguish two nearby points as distinct and separate. The maximum resolving power of the human eye when reading is 0.1 mm (100μm). We now know that the maximum resolving power of the light microscope is approximately 0.2μm, or 500× better than the human eye, and that it is dependent on the wavelength (λ) of light used for illumination as well as the numerical apertures (NA) of the objective and condenser lens systems. These are related in this equation:

$$\text{resolving power (r)} = \frac{\lambda}{NA_{obj} + NA_{cond}}$$

Examining this equation, we can see that the resolving power can be increased by decreasing the wavelength and by increasing the **numerical aperture.** Blue light affords a better resolving power than red light because its wavelength is considerably shorter. However, because the range of the visible light spectrum is rather narrow, increasing the resolution by decreasing the wavelength is of limited use. Thus, the greatest boost to the resolving power is attained by increasing the numerical aperture of the condenser and objective lens systems.

By definition, the numerical aperture equals n sin theta. The **refractive index,** n, refers to the medium employed between the objective lens and the upper slide surface as well as the medium employed between the lower slide surface and the condenser lens. With the low- and high-power objective lenses, the medium is air, which has a refractive index of 1, whereas with the oil immersion objective lens, the medium is oil, which has a refractive index of 1.25 or 1.56. Sin theta is the maximum angle formed by the light rays coming from the condenser lens and passing through the specimen into the front objective lens.

Ideally, the numerical aperture of the condenser lens should be as large as the numerical aperture of the objective lens, or the latter is reduced, resulting in reduced resolution. Practically, however, the condenser lens numerical aperture is somewhat less because the condenser iris has to be closed partially in order to avoid glare. It is also important to remember that the numerical aperture of the oil immersion objective lens depends upon the use of a dispersing medium with a refractive index greater than that of air ($n = 1$). This is achieved by using oil, which must be in contact with both the condenser lens (below the slide) and the objective lens (above the slide). When immersion oil is used on only one side of the slide, the maximum numerical aperture of the oil immersion objective is 1.25—almost the same as the refractive index of air.

Note: Oil should not be placed on the surface of the condenser lens unless your microscope contains an oil-immersion-type condenser lens and your instructor authorizes its use.

Microscopes for bacteriological use are usually equipped with three objectives: 16 mm low power (10×), 4 mm high dry power (40 to 45×), and 1.8 mm oil immersion (100×). The desired objective is rotated into place by means of a revolving nosepiece (figure 1). The millimeter number (16, 4, 1.8) refers to the **focal length** of each objective. By definition, the focal length is the distance from the principal point of focus of the objective lens to the principal point of focus of the specimen. Practically speaking, it can be said that the shorter the focal length of the objective, the shorter the **working distance** (that is, the distance between the lens and the specimen) and the larger the opening of the condenser iris diaphragm required for proper illumination (figure 2).

The power of magnification of the three objectives—10×, 45×, and 96×—is inscribed on their sides (note that these values may vary somewhat depending upon the particular manufacturer's specifications). The total magnification is obtained by multiplying the magnification of the objective lens by the magnification of the ocular eyepiece. For example, the total magnification obtained with a 4 mm objective (45×) and a 10× ocular eyepiece is 45 × 10 = 450 diameters. The highest magnification is obtained with the oil immersion objective lens. The bottom tip lens of this objective is very small and admits little light, which is why the iris diaphragm of the condenser lens must be wide open and the light conserved by means of immersion oil. The oil fills the space between the object and the objective lens so light is not lost (refer to figure 4.1 for visual explanation).

Note: If the microscope is *parfocal*, a slide that is in focus in one power will be in focus in other powers.

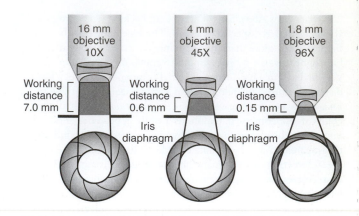

Microscope Illumination

Proper illumination is an integral part of microscopy. We cannot expect a first-class microscope to produce the best results when using a second-class illuminator. However, a first-class illuminator improves a second-class microscope almost beyond the imagination. A student microscope with only a mirror (no condenser lens) for illumination can be operated effectively by employing light from a gooseneck lamp containing a frosted or opalescent bulb. Illuminators consisting of a sheet of ground glass in front of a clear bulb are available, but they offer no advantage over a gooseneck lamp. Microscope mirrors are flat on one side and concave on the other. In the absence of a condenser lens, the concave side of the mirror should be used. Conversely, when a condenser lens is present, the flat side of the mirror should be used because condenser lenses accept only parallel rays of light and focus them on the slide.

Two or more condenser lenses are necessary for obtaining the desired numerical aperture. The Abbe condenser lens, which has a numerical aperture of 1.25, is most frequently used. The amount of light entering the objective lens is regulated by opening and closing the iris diaphragm, located between the condenser lens and the light source (figure 1). When the oil immersion objective lens is used, the iris diaphragm is opened farther than when the high- dry or low-power objective lenses are used. Focusing the light is controlled by raising or lowering the condenser lens using the condenser knob.

The mirror, condenser, and objective and ocular lenses must be kept clean to obtain optimal viewing. The ocular lenses are highly susceptible to etching from acids present in body sweat and should be cleaned after each use.

Definitions

Achromatic objective. A microscope objective lens in which the light emerging from the lens forms images practically free from prismatic colors.

Apochromatic objective. A microscope objective lens in which the light emerging from the lens forms images practically free from both spherical and chromatic aberrations.

Chromatic lens aberration. A distortion in the lens caused by the different refrangibilities of the colors in the visible spectrum.

Compound microscope. A microscope with more than one lens.

Condenser. A structure located below the microscope stage that contains a lens and iris diaphragm. It can be raised or lowered and is used for concentrating and focusing light from the illumination source on the specimen.

Focal length. The distance from the principal point of a lens to the principal point of focus of the specimen.

Figure 2 Relationship between the working distance of the objective lens and the diameter of the opening of the condenser iris diaphragm. The larger the working distance, the smaller the opening of the iris diaphragm.

Iris diaphragm. An adjustable opening that can be used to regulate the aperture of a lens.

Magnification. The ability of a microscope to increase specimen size.

Numerical aperture. A quantity that indicates the resolving power of an objective. It is numerically equal to the product of the index of refraction of the medium in front of the objective lens (n) and the sine of the angle that the most oblique light ray entering the objective lens makes with the optical axis.

Oil immersion objective lens. High-power lens (usually $100\times$) that requires the immersion of the lens into a drop of oil.

Parfocal. Having a set of objectives so mounted on the microscope that they can be interchanged without having to appreciably vary the focus.

Refractive index. The ratio of the velocity of light in the first of two media to its velocity in the second medium as it passes from one medium into another medium with a different index of refraction.

Resolution. The smallest separation that two structural forms, e.g., two adjacent cilia, must have in order to be distinguished optically as separate cilia.

Spherical lens aberration. An aberration caused by the spherical form of a lens that gives different focal lengths for central and marginal light rays.

Working distance. The distance between the tip of the objective lens when in focus and the slide specimen.

INDEX

Note: Page numbers followed by *f* and *t* indicate figures and tables, respectively.

Broth
   optical density of, 53–54
   turbidity of, 53–54
Brownian movement, 24
Budding, 140f, 141, 144
Buffer solutions, 259
   definition of, 267
   for electrophoresis, 266
Butanediol fermentation, 199, 199f

Candida, 142
   vegetative cell morphology, 140f
Candida albicans, 143f, 143t
Capsules, 31
   staining, 40–41
Carbolfuchsin stain, formula for, 295
Carbolic acid, 76, 109
Catabolite repression, 129, 131, 133–134
Catalase, definition of, 191
Catalase test, 189–190, 193, 193f
Cell lysate, 259
Cell size, measurement of, use of ocular
   micrometer for, 303–304
Centers for Disease Control and Prevention
   (CDC), 235
Cercaria, 163
   definition of, 159
   schisitosomal, 158f
Cestodes, 158
Chlamydospore, 142, 143f, 144
Chocolate agar, 190
Cholera, 235, 237
Chromatic lens aberration, 306, 308
Cilia, protozoan, 155–156
Ciliophora, 155–156
Citrate utilization, 199
Citrobacter, 210
Citrobacter freundii, 297
   identification of, 197–206
Clinical specimen, 207
Clinical unknown, identification of, 207–213
Clonorchis sinensis, 158, 162, 166
Clostridium, 41, 77
Clostridium difficile, 197
Clostridium perfringens, 239
Clostridium sporogenes, 297
Coagulase test, 182–183, 185–186
Coccidia, 156
Coccidioides immitis, 143t, 144
Coccidioidomycosis, 144
Coelomates, 158–159
   definition of, 159
Coenocytic hyphae, 139, 141f, 144
Coenzyme, 104
Coliforms, 198, 241
   definition of, 200
   membrane filter technique for, 241, 243f,
      244–246, 248–250
   multiple-tube fermentation technique for,
      239–244, 240f, 247
   in water, detection of, 238–239
Colilert test, 239
Colon, normal biota of, 197
Colony(ies), 4
   bacterial, 49
      morphology of, 207

definition of, 3
isolation of, 9, 97, 97f
Colorimeter, 67, 68f
Columella, 141f, 144, 147
Combs, for electrophoresis, 265–266, 265f, 267
Commensals, 155, 181, 183
   definition of, 159
Communicable (infectious) disease(s), 235, 251
   definition of, 251
   direct transmission of, 251
   indirect transmission of, 251
Competency, loss of, 155
Competent cells, 123
Competitive inhibition, 103, 103f
Compound (light) microscope, 17–22, 305.
   See also Microscope(s)
   definition of, 19, 308
Condenser lens, 18f, 19, 305–308
Conidia (sing., conidium), 141f, 144
Conidiophores, 141f, 144, 147
Conidiospores, 144, 147
Conjugation, gene transfer in, 115
Constitutive enzymes, 129
Contagious disease, 251
Cornmeal agar, formula for, 291
Corynebacterium, 190
Coryneform(s), 183
Counterstain, 37–38
Cryptococcus neoformans, 143t
Crystal violet (Hucker modification), formula for, 295
Culture(s). *See also* Growth; Medium (pl., media)
   microbial, 47
   mixed, 208–213
   pure, 9, 15–16, 209–213
Culture spherule, 144
Cyst(s)
   fungal, 144
   parasite, 155–156, 156f, 159, 161–162, 166
   protozoan, 135, 137, 238, 238f
Cytopathic effects (CPEs), 136, 136f, 137
Cytophaga agar, formula for, 291
Cytoplasmic membrane, and osmosis, 95, 95f

Death phase, 67, 67f
Definitive host, 156
   definition of, 159
Department of Health and Human Services
   (DHHS), 235
Dermatomycosis, 142, 144
Deuteromycetes, 140t, 142
Diarrhea, 197
Diluent, 53–54
Dilution practice problems, 299–300
Dimorphic organisms, 142, 143t, 144
Dimorphism, 142
Diphtheria, 190
Diphtheroid(s), 182–183, 190, 297
Diplococcus, 190
Direct count, 53
Dirofilaria immitis, 159, 162, 167
Disinfectant(s), 75–76, 109–110, 109t, 111–114
DNA
   amplification of, 258, 258f, 259
   bacteriophage, 171, 171f
   electrophoresis, 259, 265–270, 265f–266f

hybridization, 277–278
identification
   with polymerase chain reaction (PCR), 257–263
   with restriction enzymes, 257–263
isolation of, 261–262
light repair, 90
manipulation of, 255
methylation of, 257
microarray, 277–281, 277f
mutations, 93
   UV light and, 89
naked, 123
probe, 277
sequencing of, 259
target, 258, 258f, 259
thymine dimers, 89, 89f, 90
DNA chips. *See* Microarray(s)
DNA nucleotide sequence, 271
DNA polymerase, 258–259
DNase (deoxyribonuclease), 123, 295
Doubling time. *See* Generation time
Dry heat oven, 80–81, 85
   definition of, 78
Dugesia, 157, 157f, 162, 166, 297
Dysentery, 197
   amoebic, 155, 237
   bacterial, 237

Echinococcus, 158, 169
Ecological ingenuity, 155
Electric incinerator, 9, 10f
Electron microscope, 18
Electrophoresis, 265–270, 265f–266f
   definition of, 259, 267
Embden-Meyerhof pathway, 104. *See also*
   Glycolytic pathway
Encapsulation, 190–191
Endo broth/agar, 241
   formula for, 291
Endospore(s), 75
   definition of, 41, 90, 110
   and disinfection, 110
   heat-resistant, 77
   staining, 41–42, 41f–42f
   UV sensitivity, 89
Entamoeba gingivalis, 155
Entamoeba histolytica, 155
Enteric, definition of, 49, 200
Enteric organisms, 197
Enteric rods, 49
   identification of, 197–206
Enterobacter, 210
   colonies, 49
   identification of, 197–206
Enterobacter aerogenes, 198, 297
   detection of, 241
Enterobacteriaceae, 189, 197
Enterobius vermicularis, 159, 160f
Enterococci, 189, 191
Enterococcus faecalis, 189–190, 210, 297
Enterotube II System, 200
Enzyme(s), 223
   definition of, 215
   in immune response, 215
   inducible, 129, 130f, 131–134

Plate 1 Appearance of bacterial colonies growing on agar. The different numbers represent the different colony types. Colonies labeled as 1s, 2s, etc., are most likely the same kind of bacteria. Courtesy of Anna Oller, University of Central Missouri.

Plate 2 Appearance of *Penicillium* mold growing on agar. Courtesy of Anna Oller, University of Central Missouri.

Plate 3 An isolation streak separating *Staphylococcus* and *Micrococcus* colonies. Courtesy of Anna Oller, University of Central Missouri.

Plate 4 A negative stain showing *Bacillus* cells. Courtesy of Anna Oller, University of Central Missouri.

Plate 5 A mixed Gram stain with Gram-positive purple *Staphylococcus aureus* and Gram-negative pink *Escherichia coli* cells. *(1)* Gram-positive bacillus *(Lactobacillus)*. *(2)* Gram-negative bacillus *(Escherichia)*. *(3)* Gram-positive diplococci *(Streptococcus)*. *(4)* Gram-positive cocci in tetrads *(Micrococcus)*. *(5)* Gram-positive grapelike clusters *(Staphylococcus)*. Courtesy of Anna Oller, University of Central Missouri.

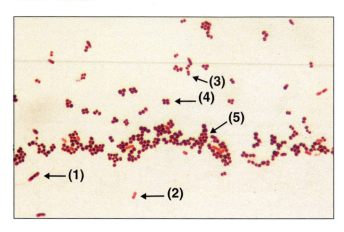

Plate 6 Acid-fast stain with methylene-blue-stained *Staphylococcus aureus* and red acid-fast-stained *Mycobacterium phlei* cells. Courtesy of Anna Oller, University of Central Missouri.

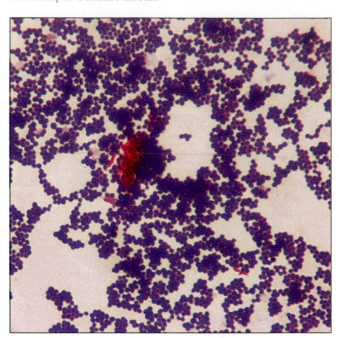

Plate 7 Capsule stain.

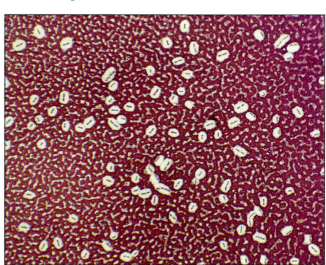

Plate 8 A positive endospore stain. *(1)* Terminal spores in *Clostridium*. *(2)* A free spore. Courtesy of Anna Oller, University of Central Missouri.

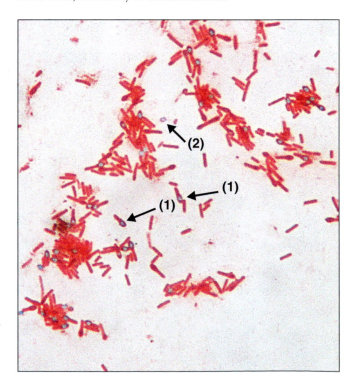

Plate 9 An eosin-methylene-blue (EMB) plate.
(1) Clear *Shigella* colonies, a negative result.
(2) A metallic-green sheen indicative of *Escherichia*, a positive result. (3) Dark purple colonies of *Enterobacter*, a positive result. Courtesy of Anna Oller, University of Central Missouri.

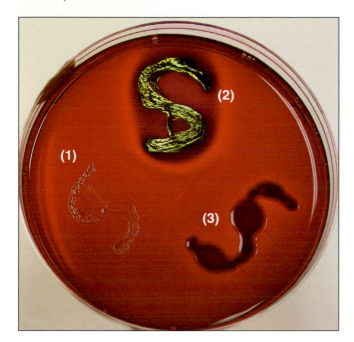

Plate 10 An autoclave. Courtesy of Anna Oller, University of Central Missouri.

Plate 12 A microscopic view at 40× of *Rhizopus nigricans*. Nonseptate hyphae lead to a sporangiospore on the right and two zygospores on the bottom left. Courtesy of Anna Oller, University of Central Missouri.

Plate 11 A starch plate after iodine addition depicting (1) a positive test with a yellow clear zone around the colony and (2) a negative test with no clear zone around the yellow colony. Courtesy of Anna Oller, University of Central Missouri.

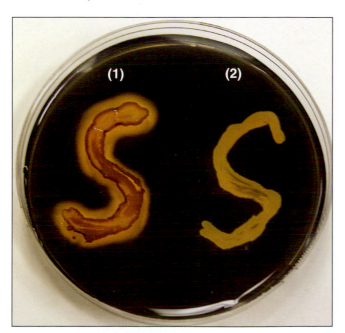

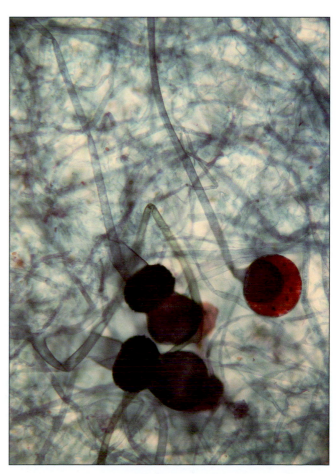

Plate 13 Ringworm caused by a dermatophyte of the beard, tinea barbae. Courtesy of CDC.

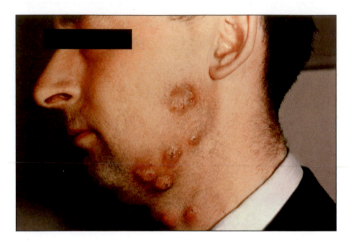

Plate 14 *Entamoeba histolytica* trophozoite. Courtesy of Anna Oller, University of Central Missouri.

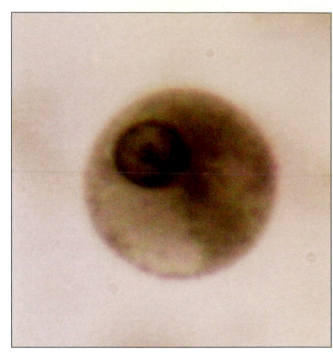

Plate 15 A *Paramecium* viewed at 40× in pond water. Courtesy of Anna Oller, University of Central Missouri.

Plate 16 A *Giardia lamblia* cyst. Courtesy of Anna Oller, University of Central Missouri.

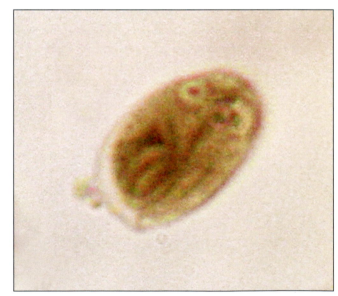

Plate 17 Flagellated *Trypanosoma* in a blood smear. Courtesy of Anna Oller, University of Central Missouri.

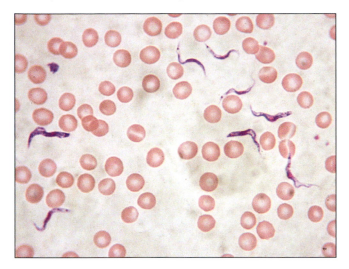

Plate 18 *Plasmodium* rings in red blood cells. Courtesy of Anna Oller, University of Central Missouri.

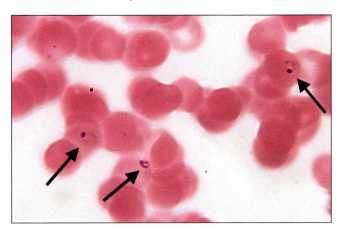

Plate 19 A *Plasmodium vivax* schizont. Courtesy of CDC.

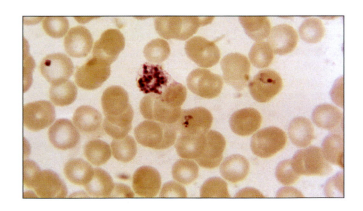

Plate 20 The liver fluke *Fasciola hepatica* viewed at 40×. Courtesy of Anna Oller, University of Central Missouri.

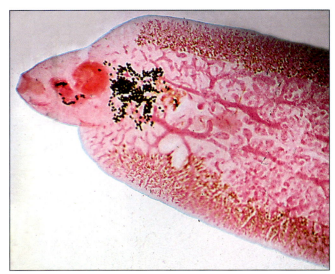

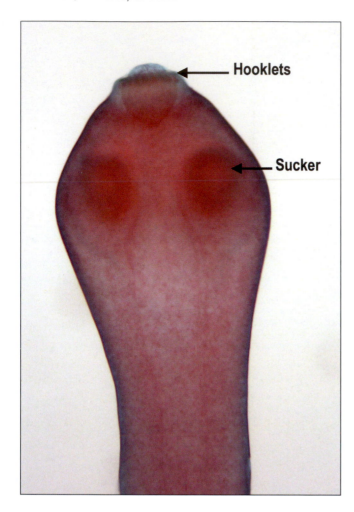

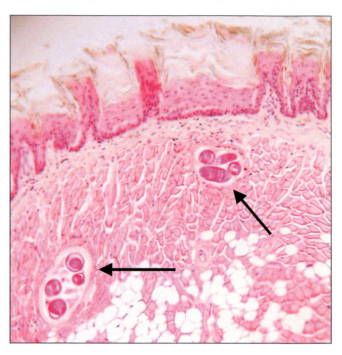

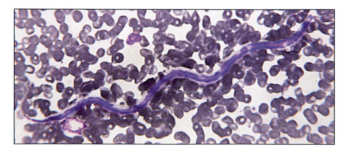

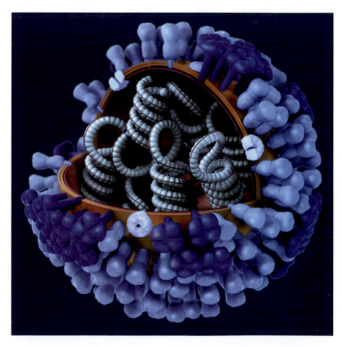

Plate 25 A mannitol salt plate with a positive *Staphylococcus aureus* culture on the left and a negative reaction by *Staphylococcus epidermidis* on the right. Courtesy of Anna Oller, University of Central Missouri.

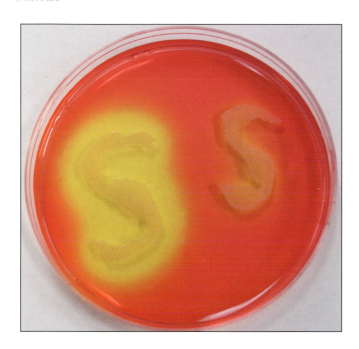

Plate 26 The coagulase test. The bottom tube contains a clump, which is a positive test indicative of *Staphylococcus aureus*, whereas the top liquid test is a negative test result.

Plate 28 Bile esculin slants depicting a positive reaction (*left*) and a negative reaction (*right*).

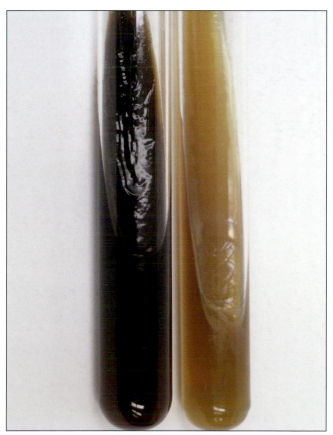

Plate 27 A blood agar plate showing colonies exhibiting α and β hemolysis. Courtesy of Anna Oller, University of Central Missouri.

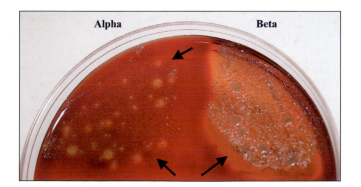

Plate 29 A strep throat infection caused by *Streptococcus pyogenes.* Courtesy of CDC.

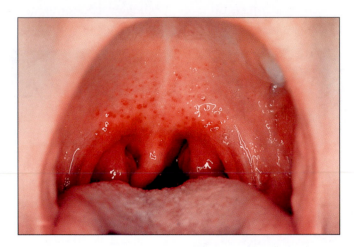

Plate 30 A catalase test. The bubbles on the left are a positive reaction, and the bacteria without bubbles show a negative reaction.

Plate 31 The oxidase test showing a positive purple reaction on the left and a negative reaction on the right.

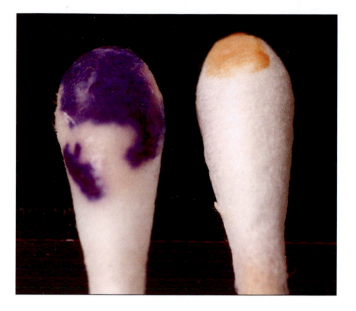

Plate 32 The colony on the *right*. *Pseudomonas aeruginosa*, produces a green pigment. Organism on *left* does not produce a pigment.

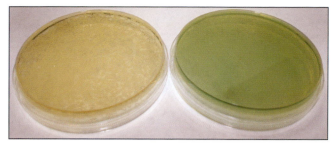

Plate 34 Glucose fermentation tubes using phenol red as the indicator. From *left* to *right*: (1) Inoculated negative control (*Pseudomonas*). (2) Acid produced (*Proteus*). (3) Acid and gas produced (*Escherichia*). Courtesy of Anna Oller, University of Central Missouri.

Plate 33 Fermentation results using Andrade's indicator. From *left* to *right*: (1) uninoculated control, (2) no change, (3) acid and gas, (4) acid. Courtesy of the University of Washington.

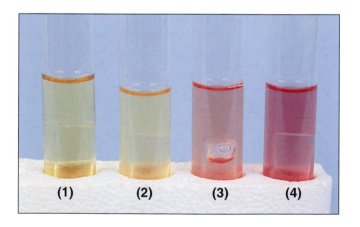

Plate 35 A MacConkey plate showing different degrees of positive results. *(1)* Yellow colony is a negative result. *(2)* Dark pink to maroon colonies are a positive result. *(3)* Positive control. Courtesy of Anna Oller, University of Central Missouri.

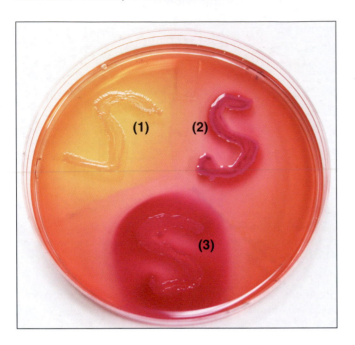

Plate 36 A Hektoen Enteric agar plate. *(1)* Clear green colonies are a negative result *(Shigella)*. *(2)* A bright orange color is a positive result *(Escherichia)*. *(3)* Green colonies with black centers are a negative result *(Salmonella)*. Courtesy of Anna Oller, University of Central Missouri.

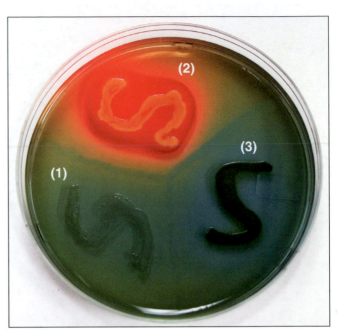

Plate 37 A *Salmonella Shigella* agar plate. *(1)* Yellow colonies are a negative result *(Shigella)*. *(2)* Yellow colonies with black centers are a negative result *(Salmonella)*. *(3)* Bright pink colonies are a positive result *(Escherichia)*. Courtesy of Anna Oller, University of Central Missouri.

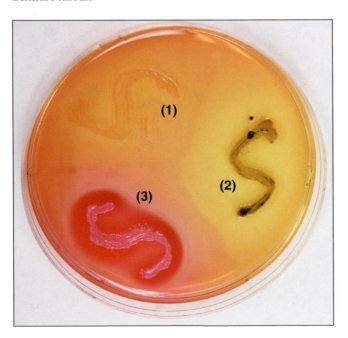

Plate 38 Triple Sugar Iron slant results. *(1)* K/K (red/red). *(2)* K/A (red/yellow). *(3)* A/A (yellow/yellow). *(4)* A/A+ (yellow/black). *(5)* A/A+ (black/black with yellow tinge). *(6)* K/A+ (red/black with yellow tinge). Courtesy of Anna Oller, University of Central Missouri.

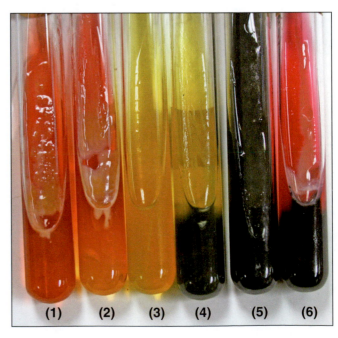

Plate 39 An XLD plate. *(1)* Red colonies with black centers are a negative result *(Salmonella)*. *(2)* Yellow colonies are a positive result *(Escherichia)*. *(3)* Clear, red colonies are a negative result *(Shigella)*. Courtesy of Anna Oller, University of Central Missouri.

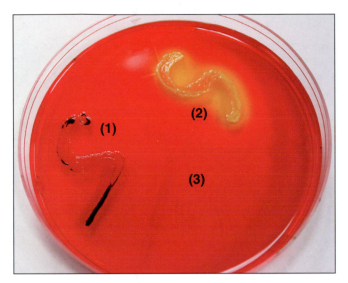

Plate 40 Methyl red test. The tube on the *right* is methyl red positive, and the tube on the *left* is methyl red negative.

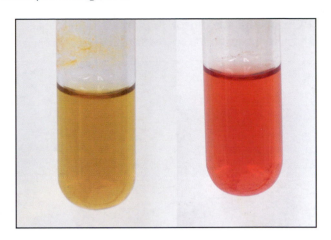

Plate 41 Voges–Proskauer tubes after reagent has been added. *(1)* A tan color is a negative result. *(2)* A brown color is a positive result. *(3)* A brick-red color is a positive result. Courtesy of Anna Oller, University of Central Missouri.

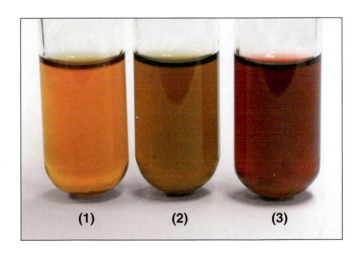

Plate 42 Citrate utilization. From *left* to *right*: *(1)* Uninoculated control. *(2)* No growth, citrate negative. *(3)* Growth, citrate positive. Courtesy of the University of Washington.

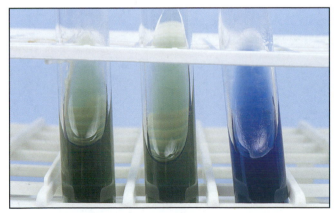

Plate 43 Urease test. From *left* to *right:*
(1) Uninoculated control. *(2)* Urease negative.
(3) Urease positive. Courtesy of the University of Washington.

Plate 44 Indol test. From *left* to *right:*
(1) Uninoculated control. *(2)* Positive for indol
formation (red). *(3)* Negative for indol formation.
Courtesy of the University of Washington.

Plate 45 Sulfur indole motility tubes after
Kovac's reagent added. *(1)* A non-motile (clear
tube), non-sulfur-producing, and indole-negative
result (*Staphylococcus*). *(2)* A motile, non-sulfur-
producing, and indole-positive result (*Escherichia*).
(3) A motile, sulfur-producing, and indole-negative
result (*Salmonella*). *(4)* A motile, sulfur-producing,
and indole-positive result (*Proteus*). Courtesy of Anna
Oller, University of Central Missouri.

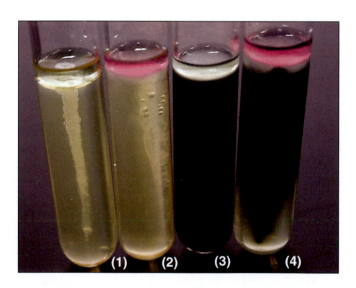

Plate 46 A phenylalanine deaminase plate after
ferric chloride addition. *(1)* Yellow is a negative
result. *(2)* Green is a positive result. Courtesy of Anna
Oller, University of Central Missouri.

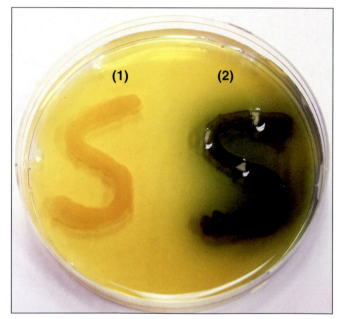

Plate 47 Rapid bacterial identification system. Enterotube II. Uninoculated tube *(top)* and inoculated tubes *(bottom)*. Courtesy of the University of Washington.

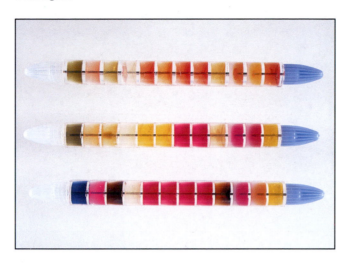

Plate 48 A blood smear stained with Wright-Giemsa. *(1)* A neutrophil. *(2)* An eosinophil. *(3)* A lymphocyte. *(4)* A platelet. *(5)* A red blood cell. Courtesy of Anna Oller, University of Central Missouri.

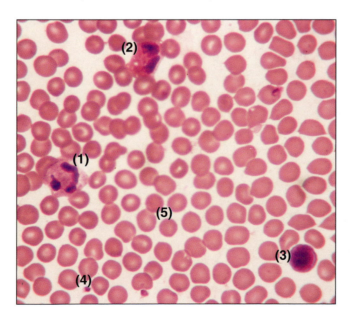

Plate 49 An ELISA 96-well plate. Courtesy of Anna
Oller, University of Central Missouri.

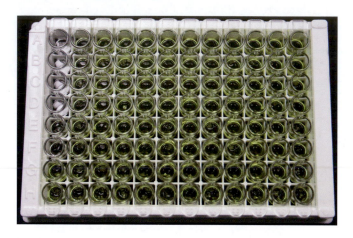

Plate 50 An electrophoresed agarose gel visualized
by ethidium bromide excitation on a UV
transilluminator. The 1 Kb ladder (standard) is on
the left, the wells are at the top, the bright band in
the middle is the amplified DNA target, and the
bands at the bottom are excess molecules. Courtesy
of Anna Oller, University of Central Missouri.

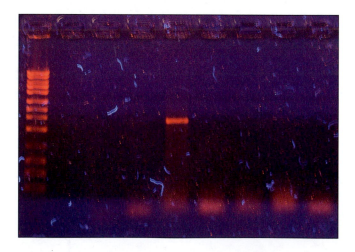